Lecture Notes in Mathematics

Volume 2394

This series reports on new developments in all areas of mathematics and their applications - quickly, informally and at a high level. Mathematical texts analysing new developments in modelling and numerical simulation are welcome. The type of material considered for publication includes:

1. Research monographs 2. Lectures on a new field or presentations of a new angle in a classical field 3. Summer schools and intensive courses on topics of current research.

Texts which are out of print but still in demand may also be considered if they fall within these categories. The timeliness of a manuscript is sometimes more important than its form, which may be preliminary or tentative. Please visit the LNM Editorial Policy (https://drive.google.com/file/d/1MOg4TbwOSokRnFJ3ZR3ciEeKs9hOnNX_/view?usp=sharing).

Titles from this series are indexed by Scopus, Web of Science, Mathematical Reviews, and zbMATH.

Władysław Wilczyński

Convergence in Measure and in Category

Springer

Władysław Wilczyński
Faculty of Mathematics and Computer
Science
University of Łódź
Łódź, Poland

ISSN 0075-8434 ISSN 1617-9692 (electronic)
Lecture Notes in Mathematics
ISBN 978-3-032-23142-0 ISBN 978-3-032-23143-7 (eBook)
https://doi.org/10.1007/978-3-032-23143-7

Mathematics Subject Classification: 28A20, 40A30

This Springer imprint is published by the registered company Springer Nature Switzerland AG
The registered company address is: Gewerbestrasse 11, 6330 Cham, Switzerland

To Urszula
To Jan Lipiński, who taught me measure theory

Preface

The notions of the Baire category and the Lebesgue measure appear frequently in the theory of real functions. The family of sets of the first category and the family of sets of Lebesgue measure zero are σ-ideals, the family of sets having the Baire property and the family of Lebesgue measurable sets are σ-algebras. The tendency to find the category analogue of a theorem concerning the Lebesgue measure (or conversely) sometimes has been very fruitful. Sierpiński [Sie34] has proved that assuming the continuum hypothesis there exists a one-to-one mapping f of the real line $\mathbb{R}$ onto itself such that $f(A)$ is a nullset if and only if A is of the first category. Erdös [Erd43] has improved the above result showing that the mapping can be such that $f = f^{-1}$. If the statement S_1 involves only the notion of nullset and the notion of set theory and S_2 is obtained from S_1 by changing "nullset" onto "set of the first category", then S_1 implies S_2 and conversely. This Sierpiński-Erdös duality principle yields numerous similarities between measure and category. The situation is different (and more interesting) when the statements involve also the notions of sets having the Baire property and the Lebesgue measurable sets. Szpilrajn [Szp34] has proved that the one-to-one mapping f of the real line $\mathbb{R}$ onto itself such that $f(E)$ is measurable if and only if E has the Baire property does not exist.

Here is the proper place to explain the title of the book. The notions of the convergence almost everywhere and the convergence in measure are widely known and play important role in the real analysis and probability. The notion of the convergence except on a set of the first category is self-explaining. To formulate the notion of the convergence in category, one can use the following version of the celebrated Riesz theorem:

Theorem 1 *A sequence $\{f_n\}_{n\in\mathbb{N}}$ of real-valued measurable functions defined on* $[0, 1]$ *converges in measure to a function f if and only if for each subsequence $\{f_{n_m}\}_{m\in\mathbb{N}}$ there exists a subsequence $\{f_{n_{m_p}}\}_{p\in\mathbb{N}}$ which converges a.e. to f (see also Theorem 2.6).*

Observe that in the above text the measure is absent. This enables to formulate the definition of the convergence in category in the form:

Definition 2 *A sequence $\{f_n\}_{n\in\mathbb{N}}$ of real-valued functions defined on* $[0, 1]$ *having the Baire property converges in category to a function f if and only if for each subsequence*

$\{f_{n_m}\}_{m\in\mathbb{N}}$ there exists a subsequence $\{f_{n_{m_p}}\}_{p\in\mathbb{N}}$ which converges to f except on a set of the first category.

In Chap. 2, one can find a general definition of the convergence with respect to the σ-ideal which covers both convergence in measure and in category (Definition 2.12). Hence, the convergence in category is a (*)-convergence with respect to the convergence except on a set of the first category similarly as the convergence in measure with respect to the convergence a.e. The notion of (*)-convergence with respect to an arbitrary prescribed convergence has been defined by Alexandrov and Urysohn [AU23] (see also [Vul67]). The great part of the excellent book [Oxt80] of John C. Oxtoby is devoted to the study of the duality between measure and Baire category. It includes a lot of theorems showing similarities or differences between measure and category. However, there is no category counterpart neither of the convergence in measure, nor of the convergence almost everywhere. The notion of Baire category connected with the convergence of a sequence of functions appears in [Oxt80] only in Chap. 8 when discussing Egorov's theorem (one can find there an ingenious construction of a sequence of continuous functions convergent everywhere but uniformly convergent only on the sets of the first category) and in Chaps. 17 and 18 (of the same book) dealing with a sequence of iterations.

The main aim of this work is to study the convergence in category and to compare it with the convergence in measure. Most frequently we shall deal with the Lebesgue measure and Baire category for the natural topology of the unit integral [0, 1]. Sometimes, we shall present more general result connected with a measurable space $(X, \mathcal{S})$ equipped with a σ-ideal $\mathcal{I} \subset \mathcal{S}$.

The content of the book is subdivided into 15 chapters (considerably independent except numbered as the first and the second). The outline of the material is as follows. For the convenience of the reader: necessary prerequisites for the reading of this book are summarized, with or without proof, in Prerequisites. Chapter 1 presents basic facts concerning the convergence almost everywhere and the convergence except on a set of the first category (denoted here the convergence $\mathcal{K}$-a.e.) and discusses theorem of Egorov from the general point of view. In Chap. 2, one can find two pseudometrics describing convergence in measure and celebrated theorem of Riesz, which is the starting point for the key definition of the convergence in category and the characterization of pairs (σ-algebra, σ-ideal) such the convergence the convergence with respect to the σ-ideal generates the Fréchet topology. Chapter 3 deals with the classical theorem of Riesz concerning the Cauchy condition in measure (so-called fundamentality in measure). It is shown how to formulate this condition in the form which enables to obtain analogous result for the convergence in category. Chapter 4 has two main subjects: the first is to describe the upper and the lower measurable (or categorial) boundaries of arbitrary functions and the second—to introduce the upper and lower limit of the sequence of measurable (or having the Baire property) functions in a reasonable way to obtain the result that such sequence of functions converges in measure (or in category) if and only if these limits coincide. In Chap. 5, the structure of measurable functions and functions having the Baire property is studied. It is proved that, for each measurable function f, there exists a function g

in the second Baire class such that $f = g$ almost everywhere and, for each function f having the Baire property, there exists a function g in the first Baire class such that $f = g$ $\mathcal{K}$-almost everywhere. Chapter 6 presents the characterization of pairs (σ-algebra, σ-ideal) for which the (*)-convergence with respect to the convergence except on a set belonging to this σ-ideal is the same as the convergence except on a set from σ-ideal. The essential role here is played by the Suslin hypothesis. Chapter 7 considers the situation when a series of real functions diverges almost everywhere or $\mathcal{K}$-almost everywhere. It is proved that if the group G of permutation of the set of positive integers is equipped with the metric d such that (G, d) is a Polish space, then the permuted series also diverges a.e or $\mathcal{K}$-a.e. for a residual set of permutations. Chapter 8 presents the classical theorems of Steinhaus confronting the convergence a.e. or divergence a.e. of series $\sum_{k=1}^{\infty} f_k$ with the convergence or divergence of series $\sum_{k=1}^{\infty} t_k f_k$, where $\{t_k\}_{k\in\mathbb{N}}$ converges to zero or to infinity. It is proved that for category some results are similar while the others are quite different. Chapter 9 deals with the sequences $\{A + x_n\}_{n\in\mathbb{N}}$ or $\{A_n + x_n\}_{n\in\mathbb{N}}$ of measurable sets or sets having the Baire property. There is studied the connection between the convergence (a.e., $\mathcal{K}$-a.e., in measure, in category) of sequences $\{\chi_{A+x_n}\}_{n\in\mathbb{N}}$, $\{\chi_{A_n+x_n}\}_{n\in\mathbb{N}}$, where $\lim_{n\to\infty} x_n = 0$, and $\left\{\frac{1}{n}\sum_{i=1}^{n} \chi_{A_i}\right\}_{n\in\mathbb{N}}$. Here also appears the difference between measure and category. The main aim of Chap. 10 is to prove that it may happen that a function of accumulation of a sequence $\{f_n\}_{n\in\mathbb{N}}$ of measurable or having the Baire property function is not measurable or has not the Baire property. As a consequence of this result, one can obtain some criterion of compactness or relative compactness. Chapter 11 deals with the sequences of measurable or having the Baire property functions defined on the unit square. It is shown that if a sequence $\{f_n(\cdot, y)\}_{n\in\mathbb{N}}$ converges a.e. (or $\mathcal{K}$-a.e.) for a.e. $y \in [0, 1]$ (or $\mathcal{K}$-a.e. $y \in [0, 1]$), then the sequence converges in both cases except on a set from $\mathcal{N}_2$ or $\mathcal{K}_2$. If the convergence is in measure or in category, the situation is quite different. Chapter 12 presents classical theorems of Mazur and Orlicz concerning the behavior of $\limsup_{n\to\infty} f(w_n x + t_n)$ where f is measurable or has the Baire property and is periodic with period 1, $\{w_n\}_{n\in\mathbb{N}}$ converges to infinity and $\{t_n\}_{n\in\mathbb{N}}$ is an arbitrary sequence. This upper limit equals to the essential upper bound of f (with respect to measure or to category) a.e. or $\mathcal{K}$-a.e. Chapter 13 includes the classical theorem of Sierpiński saying that if for each increasing sequences $\{m_k\}_{k\in\mathbb{N}}$, $\{n_k\}_{k\in\mathbb{N}}$ the sequence $\{f_{m_k,n_k}\}_{k\in\mathbb{N}}$ converges a.e., then the double sequence $\{f_{m,n}\}_{m,n\in\mathbb{N}}$ converges a.e. and its category counterpart. Chapter 14 presents the difference between measure and category concerning the set of normal numbers in the scale 2. It is shown that also the behavior of the series $\left\{\sum_{n=1}^{\infty} a_n r_n(x)\right\}_{n\in\mathbb{N}}$ where $\{r_n\}_{n\in\mathbb{N}}$ is a sequence of Rademacher, is different from the point of view of measure and category. Chapter 15 presents several topics. It is shown that the convergence in category is different from the category dual to the convergence in measure studed by Beer [Bee81], classical theorem of Fréchet concerning the compactness in measure and theorem of Trebukova saying that the convergence in measure can be changed to the convergence a.e. with the use of suitable automorphism preserving measure are proved (without counterparts for category). Some properties of basic sequences, convex and open sets in the space of measurable or having the Baire property functions are formulated.

One can observe that two important topics are left out of consideration. The first is the role of Baire theorem in the harmonic analysis. Fortunately, there is an excellent expository article by Kahane [Kah00]. The second is the method of proving existential theorems which enables to avoid burdensome constructions. Using this method, it is possible to obtain the expected result without indication any concrete object; moreover, it turns out there is a majority of such desirable objects. The method goes back to Polish mathematicians S. Saks, S. Banach, and S. Mazurkiewicz, and was developed by Bruckner and Garg ([Bru73], [BG77]).

We expect readers to have a thorough knowledge of the Lebesgue measure on the real line and the Baire category theorem and a familiarity with basic notions of set theory: cardinal and ordinal numbers, transfinite induction, trees. It is sufficient to know Chaps. 1–9, 14, and 15 of Oxtoby's book [Oxt80] and §22, Chap. 4 of Jech's book [Jec78].

I am happy to acknowledge the influence of Prof. Elżbieta Wagner-Bojakowska upon the contents of the essential part of these notes. Prof. Małgorzata Filipczak has read most of the text and has given several useful suggestions. The careful, professional, and neat typing of the manuscript by Prof. Małgorzata Terepeta (with the help by Ph.D. Aleksandra Karasińska) is greatly appreciated.

Łódź, Poland
January 2025

Władysław Wilczyński

Competing Interests The author has no competing interests to declare that are relevant to the content of this manuscript.

Prerequisites

We shall use the following denotations:

$\mathbb{R}$—the real line
$\mathbb{N}$—the set of positive integers
$\mathbb{Z}$—the set of integers
$\mathcal{S}$—σ-algebra (of subsets of a non-empty set X)
$\mathcal{I}$—σ-ideal (always $\mathcal{I} \subset \mathcal{S}$)
$\mathcal{L}$—σ-algebra of Lebesgue measurable subsets of $[0, 1]$ (or $\mathbb{R}$)
$\mathcal{N}$—σ-ideal of subsets of $[0, 1]$ (or $\mathbb{R}$) of Lebesgue measure zero
λ—one-dimensional Lebesgue measure
$\mathcal{L}_2$—σ-algebra of Lebesgue measurable subsets of $[0, 1]^2$ (or $[-1, 1]^2$ or $\mathbb{R}^2$)
$\mathcal{N}_2$—σ-ideal of subsets of $[0, 1]^2$ of Lebesgue plane measure zero
λ_2—two-dimensional Lebesgue measure
$\mathcal{B}$—σ-algebra of subsets on $[0, 1]$ (or $\mathbb{R}$) having the Baire property
$\mathcal{K}$—σ-ideal of subsets of $[0, 1]$ (or $\mathbb{R}$) of the first category
$\mathcal{B}_2$—σ-algebra of subsets on $[0, 1]^2$ (or $\mathbb{R}^2$) having the Baire property
$\mathcal{K}_2$—σ-ideal of subsets of $[0, 1]^2$ (or $\mathbb{R}^2$) of the first category
$\alpha, \beta, \gamma, \eta$—small Greek letters are used as ordinal numbers
card (A)—cardinality of the set A
ω_1—the smallest ordinal of power $\aleph_1$
A $\Subset$ B means $A \backslash B \in \mathcal{I}$ if $A, B \subset X$ or $A \backslash B$ is finite if $A, B \in \mathbb{N}$.

Definition 3 *A set A has the Baire property if it can be represented as* $A = G \triangle K$, *where G is open and K is of the first category.*

A function $f : \mathbb{R} \to \mathbb{R}$ (or $f : [0, 1] \to \mathbb{R}$) has the Baire property if the preimage of any open set has the Baire property.

Theorem 4 *The following conditions are equivalent:*

(a) $A \in \mathcal{L}$,
(b) *there exist* $B \in \mathcal{F}_\sigma$ *and* $D \in \mathcal{N}$ *such that* $A = B \cup D$,
(c) *there exist* $B \in \mathcal{G}_\delta$ *and* $D \in \mathcal{N}$ *such that* $A = B \backslash D$.

Theorem 5 *The following conditions are equivalent:*

(a) $A \in \mathcal{B}$,
(b) *there exist* $B \in \mathcal{G}_\delta$ *and* $D \in \mathcal{K}$ *such that* $A = B \cup D$,
(c) *there exist* $B \in \mathcal{F}_\sigma$ *and* $D \in \mathcal{K}$ *such that* $A = B \backslash D$.

Definition 6 *The set* $B \subset \mathbb{R}$ *is called regular open if* $B = Int(clB)$.

Theorem 7 *If* $A \in \mathcal{B}$, *then there exists a regular open set B and a set* $D \in \mathcal{K}$ *such that* $A = B \triangle D$. *Moreover, if* B_1 *is open,* $D_1 \in \mathcal{K}$ *and* $A = B_1 \triangle D_1$, *then* $B_1 \subset B$.

Definition 8 *If* $(X, \mathcal{S})$ *is a measurable space* ($X \neq \emptyset$, $S \subset 2^X$ *is a* σ*-algebra) and* $\mathcal{I} \subset \mathcal{S}$ *is a* σ*-ideal, then we shall say that some property holds* $\mathcal{I}$*-almost everywhere* ($\mathcal{I}$*-a.e.) if the set of points for which this property does not hold belongs to* $\mathcal{I}$. *If* f_1 *and* f_2 *are extended real-valued functions which are finite* $\mathcal{I}$*-a.e. on X, then we say that* f_1 *and* f_2 *are equivalent (or* $\mathcal{I}$*-equivalent) if* $\{x \in X : f_1(x) \neq f_2(x)\} \in \mathcal{I}$.

Remark 9 The above relation is indeed an equivalence relation. If g is a real-valued function which is finite $\mathcal{I}$-a.e. and defined $\mathcal{I}$-a.e. on X and we put $f(x) = g(x)$ when $g(x)$ is defined and finite and $f(x) = 0$ for remaining $x \in X$, then f is a finite real-valued function defined on X equivalent to g. This is the reason that in the text we shall always assume that all functions under consideration are defined and finite on the whole space. Observe that if g is $\mathcal{S}$-measurable, then f is also $\mathcal{S}$-measurable.

Theorem 10 (Fubini). *If* $A \in \mathcal{N}_2$, *then* $\{x \in [0, 1] : A_x \notin \mathcal{N}\} \in \mathcal{N}$, *where* $A_x = \{y \in [0, 1] : (x, y) \in A$ *is a vertical section of A.*

Remark 11 In fact it is only a small part of full version of the classical Fubini's Theorem. Obviously it holds also for horizontal sections. Similar remarks are valid for the next theorem.

Theorem 12 (Kuratowski-Ulam). *If* $A \in \mathcal{K}_2$, *then* $\{x \in [0, 1] : A_x \notin \mathcal{K}\} \in \mathcal{K}$.

Theorem 13 *There exists a set* $A \subset [0, 1]$, $A \in \mathcal{L}$ *such that* $\lambda(A \cap (a, b)) > 0$ *and* $\lambda((a, b)\backslash A) > 0)$ *for each interval* $(a, b) \subset [0, 1]$.

Proof. Let $\{(a_n, b_n)\}_{n\in\mathbb{N}}$ be a basis of the natural topology in (0,1). There exist closed nowhere dense sets D_1, E_1 such that $D_1 \notin \mathcal{N}$, $E_1 \notin \mathcal{N}$, $D_1 \cap E_1 = \emptyset$ and $D_1 \cup E_1 \subset (a_1, b_1)$. Suppose that we have constructed nowhere closed sets $D_1, \ldots, D_n$ and $E_1, \ldots, E_n$ of positive measure such that $\left(\bigcup_{i=1}^{n} D_i\right) \cap \left(\bigcup_{i=1}^{n} E_i\right) = \emptyset$ and $D_i \cup E_i \subset (a_i, b_i)$ for $i = 1, 2, \ldots, n$. To construct D_{n+1} and E_{n+1} observe that the set $B = (a_{n+1}, b_{n+1})\backslash\left(\bigcup_{i=1}^{n}(D_i \cup E_i)\right)$ is open and non-empty. Let (c, d) be a component of B. There exist closed nowhere dense sets D_{n+1}, E_{n+1} such that $D_{n+1} \notin \mathcal{N}$, $E_{n+1} \notin \mathcal{N}$, $D_{n+1} \cap E_{n+1} = \emptyset$, $D_{n+1} \cup E_{n+1} \subset (c, d) \subset (a_{n+1}, b_{n+1})$. The set $A = \bigcup_{n=1}^{\infty} D_n$ fulfills all requirements. □

Theorem 14 *If a real function defined on* [0,1] *has the Baire property, then there exists a residual set* $A \subset [0, 1]$ *such that* $f|_A$ *is continuous.*

Remark 15 The situation is different when f is measurable (compare Lusin's theorem—Theorem 5.3).

Contents

Chapter 1
Convergence a.e. and $\mathcal{K}$-a.e.

In this section it is shown that measurability is preserved under the operation of upper and lower limits of sequences of functions. Also the interrelation between $\mathcal{I}$-a.e. convergence and uniform convergence on big subsets is studied in abstract setting and is compared with the cases of Lebesgue measurability or the Baire property of functions.

Suppose that $(X, \mathcal{S})$ is a measurable space and $\mathcal{I} \subset \mathcal{S}$—a σ-ideal.

Theorem 1.1 *Let $\{f_n\}_{n\in\mathbb{N}}$ be a sequence of $\mathcal{S}$-measurable real-valued functions defined on X. Then* $\limsup_{n\to\infty} f_n$ *and* $\liminf_{n\to\infty} f_n$ *are $\mathcal{S}$-measurable.*

Proof For each $a \in \mathbb{R}$ we have

$$\{x \in X\colon \sup_n f_n(x) > a\} = \bigcup_{n=1}^{\infty}\{x \in X\colon f_n(x) > a\}$$

and

$$\{x \in X\colon \inf_n f_n(x) < a\} = \bigcup_{n=1}^{\infty}\{x \in X\colon f_n(x) < a\},$$

so $\sup_n f_n$ and $\inf_n f_n$ are $\mathcal{S}$-measurable. Since

$$\limsup_{n\to\infty} f_n = \lim_{n\to\infty}\left(\sup_{i\geq n} f_i(x)\right) = \inf_{n\in\mathbb{N}}\left(\sup_{i\geq n} f_i(x)\right)$$

and

$$\liminf_{n\to\infty} f_n = \lim_{n\to\infty}\left(\inf_{i\geq n} f_i(x)\right) = \sup_{n\in\mathbb{N}}\left(\inf_{i\geq n} f_i(x)\right),$$

upper and lower limits of $\{f_n\}_{n\in\mathbb{N}}$ are $\mathcal{S}$-measurable. □

W. Wilczyński, *Convergence in Measure and in Category*, Lecture Notes in Mathematics 2394, https://doi.org/10.1007/978-3-032-23143-7_1

Corollary 1.2 *If $\{f_n\}_{n\in\mathbb{N}}$ is a sequence of $\mathcal{S}$-measurable real-valued functions defined on X such that $\lim_{n\to\infty} f_n = f$ $\mathcal{I}$-a.e., then f is $\mathcal{S}$-measurable.*

Remark 1.3 Suppose that all functions below are real-valued measurable functions defined on $[0, 1]$. If $f = g$ a.e., then we say that these functions are equivalent ($f \sim g$). If $f_n \sim h_n$ for $n \in \mathbb{N}$, $f \sim g$ and $\lim_{n\to infty} f_n = f$ a.e., then also $\lim_{n\to infty} h_n = g$ a.e.

If $[f] = \{g\colon f \sim g\}$ denotes the equivalence class of a function f, then we can write $\lim_{n\to infty}[f_n] = [f]$ a.e. if and only if $\lim_{n\to infty} f_n = f$ a.e.

To simplify the denotations we shall formulate all statements in terms of functions rather than in terms of equivalence classes.

We shall start with Egorov's theorem and investigate the role of measurability.

Convergence a.e.

Theorem 1.4 (Egorov, see *[Ego11]*) *If $\{f_n\}_{n\in\mathbb{N}}$ is a sequence of real-valued Lebesgue measurable functions defined on $[0, 1]$ convergent a.e. to a.e. finite function f, then for each $\varepsilon > 0$ there exists a set $E_\varepsilon \subset [0, 1]$ such that $\lambda([0, 1] \setminus E_\varepsilon) < \varepsilon$ and the sequence $\{f_n\}_{n\in\mathbb{N}}$ converges uniformly to f on E_ε.*

Proof For each $n, k \in \mathbb{N}$ let

$$E_{n,k} = \left\{x \in [0, 1]\colon |f_m(x) - f(x)| < \frac{1}{k} \text{ for each } m \geq n\right\}$$
$$= \bigcap_{m=n}^{\infty} \left\{x \in [0, 1]\colon |f_m(x) - f(x)| < \frac{1}{k}\right\}.$$

These sets are measurable, since f is also measurable. If $\lim_{n\to\infty} f_n(x) = f(x)$, then for each $k \in \mathbb{N}$ and for sufficiently big n we have $x \in E_{n,k}$. Hence

$$\{x \in [0, 1]\colon \lim_{n\to\infty} f_n(x) = f(x)\} \subset \bigcup_{n=1}^{\infty} E_{n,k}$$

for each $k \in \mathbb{N}$ and from the assumption and the fact that the sequence $\{E_{n,k}\}_{n\in\mathbb{N}}$ is ascending it follows that $\lim_{n\to\infty} \lambda(E_{n,k}) = 1$. Let $\varepsilon > 0$. For each $k \in \mathbb{N}$ there exists $n_k \in \mathbb{N}$ such that $\lambda\left([0, 1] \setminus E_{n_k,k}\right) < \varepsilon \cdot 2^{-k}$. Put $E_\varepsilon = \bigcap_{k=1}^{\infty} E_{n_k,k}$. Then

$$\lambda([0, 1] \setminus E_\varepsilon) \leq \sum_{k=1}^{\infty} \lambda\left([0, 1] \setminus E_{n,k}\right) < \sum_{k=1}^{\infty} \varepsilon \cdot 2^{-k} = \varepsilon.$$

From the definition of the sets $E_{n,k}$ it follows that, for $m \geq n_k$, $|f_m(x) - f(x)| < \frac{1}{k}$ for each $x \in E_{n,k}$, so for each $x \in E_\varepsilon$, which means that $\{f_n\}_{n\in\mathbb{N}}$ converges to f uniformly on E_ε. □

N. Lusin in [Lus34] has proved an interesting theorem from which the theorem of Egorov follows immediately. To take into consideration also the case when limits assume infinite values, we shall agree that $-\infty - \varepsilon = -\infty$, $-\infty + \varepsilon = -\frac{1}{\varepsilon}$, $\infty - \varepsilon = \frac{1}{\varepsilon}$, $\infty + \varepsilon = \infty$ for $\varepsilon > 0$.

We shall say that a sequence $\{f_n\}_{n\in\mathbb{N}}$ of real valued functions defined on $[0, 1]$ uniformly oscillates on a set $E \subset [0, 1]$ if and only if for each $\varepsilon > 0$ there exists $n_0 \in \mathbb{N}$ such that for each $n \geq n_0$ and $x \in E$

$$\liminf_{n\to\infty} f_n(x) - \varepsilon < f_n(x) < \limsup_{n\to\infty} f_n(x) + \varepsilon.$$

Theorem 1.5 *Each sequence $\{f_n\}_{n\in\mathbb{N}}$ of real-valued measurable functions defined on $[0, 1]$ uniformly oscillates on a set $E \subset [0, 1]$, the measure of which is arbitrarily closed to 1.*

Proof Let

$$E_{n,k} = \left\{x \in [0, 1]\colon \liminf_{n\to\infty} f_n(x) - \frac{1}{k} < f_m(x) < \limsup_{n\to\infty} f_n(x) + \frac{1}{k} \text{ for each } m \geq n\right\}$$

These sets are measurable, since $\liminf_{n\to\infty} f_n(x)$ and $\limsup_{n\to\infty} f_n(x)$ are measurable. Observe that $[0, 1] = \bigcup_{n=1}^{\infty} E_{n,k}$ for each $k \in \mathbb{N}$ and that the sequence $\{E_{n,k}\}_{k\in\mathbb{N}}$ is ascending for each $k \in \mathbb{N}$. The rest of the proof goes similarly as in the Egorov's theorem. □

Remark 1.6 Without the assumption of measurability the theorem does not hold.

Indeed, let $\{E_m\}_{m\in\mathbb{N}}$ be a sequence of subsets of $[0, 1]$ such that $\lambda^*(E_m) = 1$ for each $m \in \mathbb{N}$, $\bigcup_{m=1}^{\infty} E_m$ and $E_{m_1} \cap E_{m_2} = \varnothing$ for $m_1 \neq m_2$. Put $A_n = \bigcup_{m=n}^{\infty} E_m$ and $f_n = \chi_{A_n}$. It is easy to see that $\lim_{n\to\infty} f_n(x) = 0$ for each $x \in [0, 1]$. Simultaneously, if $A \subset [0, 1]$ is a measurable set such that $\{f_n\}_{n\in\mathbb{N}}$ converges uniformly on A to 0, then there exists $n_0 \in \mathbb{N}$ such that $A \subset \bigcup_{m=1}^{n_0} E_m$, hence $\lambda(A) \leq \lambda_*\left(\bigcup_{m=1}^{n_0} E_m\right) = 0$ (compare [Sie17, Sie27]).

Remark 1.7 If instead of a sequence $\{f_n\}_{n\in\mathbb{N}}$ we have a family $\{f_y\}_{y\in[1,\infty)}$ of measurable functions such that for $x \in [0, 1]$, $\lim_{y\to\infty} f_y(x) = f(x)$, then the analogous theorem does not hold.

Indeed, let $\{E_m\}_{m\in\mathbb{N}}$ be a sequence of sets from previous remark. Put $A = \bigcup_{m=1}^{\infty} A_m$, where $A_m = \{(x, m + x)\colon x \in E_m\}$ and $f_y(x) = \chi_A(x, y)$. For each $y \in [1, \infty)$ f_y is measurable as a function equal to 0 except at most one point, moreover $\lim_{y\to\infty} f_y(x) = 0$ for each $x \in [0, 1]$. If $A \subset [0, 1]$ is a measurable set on which $\{f_y\}_{y\in[1,\infty)}$ converges uniformly, then $\lambda(A) = 0$ (the proof of the previous remark works) (compare [Wal77]).

Sets of uniform convergence in Remark 1.6 are small in the sense of measure, but can be big in the sense of cardinality. W. Sierpiński with the aid of Continuum Hypothesis has proved that without the assumption that considered functions are measurable the uniform convergence can hold only on countable sets.

Theorem 1.8 *([Sie28]) There exists a sequence* $\{f_n\}_{n\in\mathbb{N}}$ *of real-valued functions defined on* $[0, 1]$ *such that* $\lim_{n\to\infty} f_n(x) = 0$ *for each* $x \in [0, 1]$ *and if* $E \subset [0, 1]$ *is the set such that* $\{f_{n|_E}\}_{n\in\mathbb{N}}$ *converges uniformly, then* $\text{card}(E) \leq \aleph_0$.

Proof Let $\mathcal{S}$ be the set of all increasing sequences of positive integers. Since $\text{card}(S) = \mathfrak{c}$, there exists a transfinite sequence $\{s_\beta\}_{\beta<\omega_1}$ consisting of all elements of $\mathcal{S}$. We shall construct a transfinite sequence $\{u_\alpha\}_{\alpha<\omega_1}$ of elements of $\mathcal{S}$ in the following way:

Put $u_0 = s_0$. Suppose that we have defined $u_\xi = \{n_1^\xi, n_2^\xi, \ldots\}$ for $\xi < \alpha$, where $\alpha < \omega_1$. Since the set $\{u_\xi : \xi < \alpha\}$ is at most denumerable, there exists a sequence $u = \{n_1, n_2, \ldots\}$ fulfilling the following condition (*): for each $\xi < \alpha$ there exists $q = q(\xi)$ such that $n_k^\xi < n_k$ for $k \geq q(\xi)$. Indeed, if $\{\xi_1, \xi_2, \ldots\}$ is a sequence of all ordinal numbers less than α, it is sufficient to put $n_k = n_k^{\xi_1} + n_k^{\xi_2} + \cdots n_k^{\xi_k}$ for $k \in \mathbb{N}$ and $q(\xi_i) = i + 1$ for $i \in \mathbb{N}$. Let $u_\alpha = \{n_1, n_2, \ldots\}$ be the first element of $\{s_\beta\}_{\beta<\omega_1}$ fulfilling (*). The sequence $\{u_\alpha\}_{\alpha<\omega_1}$ has been defined.

Let $\{x_\alpha\}_{\alpha<\omega_1}$ be a transfinite sequence consisting of all numbers from $[0, 1]$. We shall define a sequence $\{f_n\}_{n\in\mathbb{N}}$ of functions in the following way: If $x \in [0, 1]$, then there exists $\alpha < \omega_1$ such that $x = x_\alpha$. We want to define $f_n(x)$ for $n \in \mathbb{N}$. If n is an element of $u_\alpha = \{n_1^\alpha, n_2^\alpha, \ldots\}$, then there exists $k \in \mathbb{N}$ such that $n = n_k^\alpha$. In this case put $f_n(x) = \frac{1}{k}$. If n does not belong to u_α, put $f_n(x) = 0$. It is obvious that $\lim_{n\to\infty} f_n(x) = 0$ for $x \in [0, 1]$, since all sequences u_α are increasing.

Suppose now that $E \subset [0, 1]$ is non-denumerable set (so $\text{card}(E) = \mathfrak{c}$)) on which the sequence $\{f_n\}_{n\in\mathbb{N}}$ is uniformly convergent. There exists an increasing sequence $\{n_k\}_{k\in\mathbb{N}}$ of positive integers such that $|f_n(x)| < \frac{1}{k}$ for $n \geq n_k$ and for each $x \in E$. Let $\{\alpha_\xi\}_{\xi<\omega_1}$ be an increasing transfinite sequence of ordinal numbers such that $E = \{x_{\alpha_\xi} : \xi < \omega_1\}$. Obviously $\{\alpha_\xi : \xi < \omega_1\}$ is cofinal with $\{\beta : \beta < \omega_1\}$. If there exist $k \in \mathbb{N}$ and $\xi < \omega_1$ for which $n_k^{\alpha_\xi} \geq n_k$, then for $x = x_{\alpha_\xi}$ and $n = n_k^{\alpha_\xi}$ we have $f_n(x) = \frac{1}{k}$, but for $n \geq n_k$ it should be $f_n(x) < \frac{1}{k}$ - a contradiction. Hence $n_k^{\alpha_\xi} < n_k$ for each $\xi < \omega_1$ and $k \in \mathbb{N}$. Let $\mu < \omega_1$ be an ordinal number for which $s_\mu = (n_1, n_2, \ldots)$. There exists $\xi < \omega_1$ such that $\alpha_\xi > \mu$. From the construction of u_{α_ξ} it follows that it cannot be $n_k^{\alpha_\xi} < n_k$ for each $k \in \mathbb{N}$, a contradiction. □

Observe that the theorem of Egorov can be reformulated in the following way:

Theorem 1.9 *If* $\{f_n\}_{n\in\mathbb{N}}$ *is a sequence of real-valued Lebesgue measurable functions defined on* $[0, 1]$ *convergent a.e. to a.e. finite function* f*, then there exists a sequence* $\{A_m\}_{m\in\mathbb{N}}$ *of measurable sets such that* $\lambda\left([0, 1] \setminus \bigcup_{m=1}^{\infty} A_m\right) = 0$ *and* $\{f_n\}_{n\in\mathbb{N}}$ *converges to uniformly on each* A_m.

Proof It is sufficient to take a sequence $\{\varepsilon_m\}_{m\in\mathbb{N}} = \left\{\frac{1}{m}\right\}_{m\in\mathbb{N}}$ and put $A_m = E_{\varepsilon_m}$. □

Obviously the converse theorem is also true. S. J. Taylor has presented in [Tay60] another form of the Egorov theorem, in which it is not necessary to assume that the functions f_n are defined on a set of finite measure.

Theorem 1.10 *(see [Tay60]) If $E \subset \mathbb{R}$ is a Lebesgue measurable set and $\{f_n\}_{n\in\mathbb{N}}$ is a sequence of measurable functions defined on E convergent a.e. to a.e. finite function f, then there exists a non-increasing sequence $\{\delta_n\}_{n\in\mathbb{N}}$ convergent to 0 and a subset $S \subset E$ such that $\lambda(E \setminus S) = 0$ and $\lim_{n\to\infty} \delta_n^{-1}|f_n(x) - f(x)| = 0$ for each $x \in S$.*

Proof The set E can be represented in the from $E = \bigcup_{k=1}^{\infty} E_k$, where each E_k is measurable and $\lambda(E_k) < \infty$. From the theorem of Egorov it follows that for each $k \in \mathbb{N}$ there exists a sequence $\{E_{k,m}\}_{m\in\mathbb{N}}$ of measurable subsets of E_k such that $\lambda(E_k \setminus E_{k,m}) < 2^{-m}$ and $\{f_n\}_{n\in\mathbb{N}}$ converges uniformly to f on $E_{k,m}$. Put $S = \bigcup_{k=1}^{\infty}\bigcup_{m=1}^{\infty} E_{k,m}$. Since $\lambda\left(E_k \setminus \bigcup_{m=1}^{\infty} E_{k,m}\right) = 0$, we have $\lambda(E \setminus S) = 0$. For fixed k, m let $\{n_{k,m,r}\}_{r\in\mathbb{N}}$ be an increasing sequence of positive integers such that $|f_n(x) - f(x)| < \frac{1}{r+1}$ for each $n \geq n_{k,m,r}$ and $x \in E_{k,m}$. Let $\{n_p\}_{p\in\mathbb{N}}$ be an increasing sequence of positive integers such that $\lim_{p\to\infty} \frac{n_p}{n_{k,m,p}} = \infty$ for each fixed $k, m \in \mathbb{N}$. (For example, take $n_p = p \cdot \max\{n_{i,j,p} : i \leq p, j \leq p\}$. Then for each $p \geq \max\{k, m\}$ we have $n_p \geq p \cdot n_{k,m,p}$). Define

$$\delta_n = \begin{cases} 1 & \text{for } 1 \leq n \leq n_1 \\ p^{-\frac{1}{2}} & \text{for } n_{p-1} + 1 \leq n \leq n_p,\ p \geq 2. \end{cases}$$

A sequence $\{\delta_n\}_{n\in\mathbb{N}}$ is non-increasing and $\lim_{n\to\infty} \delta_n = 0$. If $x \in S$, then there exist $k, m \in \mathbb{N}$ such that $x \in E_{k,m}$. There exists $p_0 \in \mathbb{N}$ such that $n_p > n_{k,m,p}$ for $p \geq p_0$. Hence $|f_n(x) - f(x)| < \frac{1}{p+1} < \delta_n^2$ for $n \geq n_{p_0}$. That means that $\lim_{n\to\infty} \delta_n^{-1}|f_n(x) - f(x)| = 0$ for each $x \in S$. □

E. Zakon in [Zak64] has proved another version of the theorem of Egorov for real functions defined on the whole real line $\mathbb{R}$ (in fact, in more general σ-finite measure spaces). A sequence of real-valued functions $\{f_n\}_{n\in\mathbb{N}}$ is said to converge subuniformly to a function f on a set $A \subset \mathbb{R}$ if each point $x \in \mathbb{R}$ has an open neighbourhood G_x such that $\{f_n\}_{n\in\mathbb{N}}$ converges uniformly to f on $A \cap G_x$.

Theorem 1.11 *([Zak64]) If $\{f_n\}_{n\in\mathbb{N}}$ is a sequence of real-valued measurable functions defined on $\mathbb{R}$ convergent a.e. to a.e. finite function f, then for each $\varepsilon > 0$ there exists a measurable set $A \subset \mathbb{R}$ such that $\lambda(\mathbb{R} \setminus A) < \varepsilon$ and $\{f_n\}_{n\in\mathbb{N}}$ converges f to subuniformly on A.*

Proof Let $\{G_k\}_{k\in\mathbb{N}}$ be the sequence of open sets of finite Lebesgue measure such that $\mathbb{R} = \bigcup_{k=1}^{\infty} G_k$. Put $D_1 = G_1$ and $D_k = G_k \setminus \bigcup_{i=1}^{k-1} G_i$ for $k \geq 2$. Then $\mathbb{R} = \bigcup_{k=1}^{\infty} D_k$, all $D_k's$ are measurable of finite measure. Take $\varepsilon > 0$. For each $k \in \mathbb{N}$ there exists a measurable set $A_k \subset D_k$ such that $\lambda(D_k \setminus A_k) < \frac{\varepsilon}{2^k}$ and $\{f_n\}_{n\in\mathbb{N}}$ converges to f uniformly on A_k. Put $A = \bigcup_{k=1}^{\infty} A_k$. Obviously A is measurable and $\lambda(\mathbb{R} \setminus A) < \varepsilon$. For each $x \in \mathbb{R}$ there exists $k_x \in \mathbb{N}$ such that $x \in G_{k_x}$. From the definition it follows that all sets D_k witch $k > k_x$ are disjoint with G_{k_x}. Hence all sets A_k with $k > k_x$

are disjoint with G_{k_x}, so $G_{k_x} \cap \bigcup_{k=k_x+1}^{\infty} A_k = \varnothing$ and $A \cap G_{k_x} \subset \bigcup_{k=1}^{k_x} A_k$. Since $\{f_n\}_{n\in\mathbb{N}}$ converges to f uniformly on each A_k, $k = 1, \ldots, k_x$, we have also that $\{f_n\}_{n\in\mathbb{N}}$ converges to f uniformly on $\bigcup_{k=1}^{k_x} A_k$ and on $A \cap G_{k_x}$ as required. □

For the convergence $\mathcal{K}$-a.e. (except on a set of first category) the situation is different. J. C. Oxtoby in [Oxt80], p. 38, has presented the following example: let $\Phi(x) = 2x$ on $[0, \frac{1}{2}]$, $\Phi(x) = 2 - 2x$ on $[\frac{1}{2}, 1]$, and $\Phi(x) = 0$ on $\mathbb{R} \setminus [0, 1]$. Put $f_n(x) = \sum_{i=1}^{\infty} 2^{-i} \cdot \Phi\left(2^n(x - r_i)\right)$, where $\{r_i\}_{i\in\mathbb{N}}$ is a set dense in $\mathbb{R}$. Then each f_n is continuous as a sum of a uniformly convergent series of continuous functions, $\lim_{n\to\infty} f_n(x) = 0$ everywhere in $\mathbb{R}$ and each set on which the sequence $\{f_n\}_{n\in\mathbb{N}}$ converges uniformly is (even) nowhere dense.

A sequence $\{f_n\}_{n\in\mathbb{N}}$ defined above is not monotone. In Chapter 8 one can find another example in which $f_{n+1} \le f_n$ for $n \in \mathbb{N}$, however functions are not continuous (and cannot be).

Functions in Sierpiński's Theorem 1.8 cannot be measurable, but it is not immediate that they have not the Baire property. It will follow from the following theorem:

Theorem 1.12 *Let $\{f_n\}_{n\in\mathbb{N}}$ be a sequence of real-valued functions having the Baire property defined on $[0, 1]$. Suppose that $\{f_n\}_{n\in\mathbb{N}}$ converges to a function $f_0\colon [0, 1] \to \mathbb{R}$ except on a set $B \subset [0, 1]$ of the first category. Then there exists a set $D \subset [0, 1]$ such that $\operatorname{card}(D) = \mathfrak{c}$ and the sequence $\{f_n|_D\}_{n\in\mathbb{N}}$ converges uniformly to $f_0|_D$.*

Proof The function f_0 also has the Baire property (Corollary 1.2), so for each $n \in \mathbb{N} \cup \{0\}$ there exists a set E_n of the first category such that $f_n|_{[0,1]\setminus E_n}$ is continuous. Put $E = B \cup \bigcup_{n=0}^{\infty} E_n$ and let $A \subset [0, 1]$ be a set such that $E \subset A$, $A \in \mathcal{F}_\sigma$ and A is of the first category. Then $[0, 1] \setminus A$ is a $\mathcal{G}_\delta$ set residual on $[0, 1]$. From the theorem of Alexandrov–Hausdorff ([Kur24], t.I, p. 447) it follows that there exists a non-empty perfect set $F \subset [0, 1] \setminus A$. The sequence $\{f_n|_F\}_{n\in\mathbb{N}}$ consists of continuous functions and converges pointwise to $f_0|_F$. Put

$$g_n(x) = \begin{cases} f_n(x) & \text{for } x \in F \\ 0 & \text{for } x \in [0, 1] \setminus F \end{cases}$$

for $n \in \mathbb{N} \cup \{0\}$. Then $\{g_n\}_{n\in\mathbb{N}}$ is a sequence of measurable functions convergent pointwise to g_0. Let $h\colon [0, 1] \to [0, 1]$ be a homeomorphism such that $\lambda(h(F)) = a > 0$. Let $k_n(z) = g_n\left(h^{-1}(z)\right)$ for $z \in [0, 1]$ and $n \in \mathbb{N} \cup \{0\}$. Then $\{k_n\}_{n\in\mathbb{N}}$ is a sequence of measurable functions convergent pointwise to k_0. Put $\varepsilon = \frac{a}{2}$. From the theorem of Egorov it follows that there exists a closed set $D_\varepsilon \subset [0, 1]$ such that $\lambda\left([0, 1] \setminus D_\varepsilon\right) < \varepsilon$ and $\{k_n|_{D_\varepsilon}\}_{n\in\mathbb{N}}$ uniformly converges to $k_0|_{D_\varepsilon}$. But $\lambda\left(h(F) \cap D_\varepsilon\right) > \frac{a}{2}$, so $\operatorname{card}(h(F) \cap D_\varepsilon) = \mathfrak{c}$ and also $\operatorname{card}\left(h^{-1}\left(h(F) \cap D_\varepsilon\right)\right) = \mathfrak{c}$. Denote $D = h^{-1}\left(h(F) \cap D_\varepsilon\right)$ and observe that $\{g_n|_D\}_{n\in\mathbb{N}}$ converges uniformly to $g_0|_D$ and $D \subset F$, so finally $\{f_n|_D\}_{n\in\mathbb{N}}$ converges uniformly to $f_0|_D$. □

Recall that in a metric space if a sequence $\{p_n\}_{n\in\mathbb{N}}$ does not converge to p_0, then there exists a subsequence $\{p_{n_m}\}_{m\in\mathbb{N}}$ no subsequence of which converges to p_0.

Theorem 1.13 *([Vai64], p. 233) The almost everywhere convergence of measurable functions defined on* $[0, 1]$ *cannot be induced by a metric.*

Proof Consider the following sequence $\{f_n\}_{n\in\mathbb{N}}$ of functions: each positive integer n can be uniquely represented in the from $n = 2^k + i$, where $k \in \mathbb{N} \cup \{0\}$ and $0 \leq i < 2^k$. Put $f_n = \chi_{\left[\frac{i}{2^k}, \frac{i+1}{2^k}\right]}$. Then a sequence $\{f_n\}_{n\in\mathbb{N}}$ does not converge to $f_0 \equiv 0$, but for each subsequence $\{f_{n_m}\}_{m\in\mathbb{N}}$ of $\{f_n\}_{n\in\mathbb{N}}$ there exists a subsequence $\{f_{n_{m_p}}\}_{p\in\mathbb{N}}$ such that $\lim_{p\to\infty} f_{n_{m_p}} = f_0$ except at most one-point set. □

The same sequence suffices to prove

Theorem 1.14 *The $\mathcal{K}$-almost everywhere convergence of functions defined on* $[0, 1]$ *cannot be induced by a metric.*

In the family $\mathcal{M}$ of Lebesgue measurable real-valued functions defined on $[0, 1]$ we can define the closure operation assuming that f belongs to the closure of the set $A \subset \mathcal{M}$ if and only if there exists a sequence $\{f_n\}_{n\in\mathbb{N}}$ of functions from A such that $\lim_{n\to\infty} f_n = f$ a.e. We shall show that the closure operation fulfills all axioms of Kuratowski, so it generates a topology.

Theorem 1.15 *For each $A \subset \mathcal{M}$: $A \subset \overline{A}$ and $\overline{A} = \overline{\overline{A}}$; for each sets $A, B \subset \mathcal{M}$: $\overline{A \cup B} = \overline{A} \cup \overline{B}$ and $\overline{\varnothing} = \varnothing$.*

Proof It is easy to see that $A \subset \overline{A}$, $\overline{A \cup B} = \overline{A} \cup \overline{B}$ and $\overline{\varnothing} = \varnothing$. Let $A \subset \mathcal{M}$. It is obvious that $\overline{A} \subset \overline{\overline{A}}$. Take $f \in \overline{\overline{A}}$. There exists a sequence $\{f_n\}_{n\in\mathbb{N}}$ of functions from $\overline{A}$ such that $f = \lim_{n\to\infty} f_n$ a.e. and for each $n \in \mathbb{N}$ there exists a sequence $\{f_{n,k}\}_{k\in\mathbb{N}}$ from A such that $f_n = \lim_{k\to\infty} f_{n,k}$ a.e. From the theorem of Egorov (Theorem 1.5) for each $n \in \mathbb{N}$ there exists a measurable set $E_n \subset [0, 1]$ such that $\lambda(E_n) > 1 - \frac{1}{2^n}$ and $\{f_{n,k}\}_{k\in\mathbb{N}}$ converges uniformly to f_n on the set E_n. Hence there exists $k_n \in \mathbb{N}$ such that $\left|f_n(x) - f_{n_k}\right| < \frac{1}{n}$ for each $x \in E_n$. Let $B \subset [0, 1]$ be a measurable set such that $\lambda(B) = 1$ and $f(x) = \lim_{n\to\infty} f_n(x)$ for each $x \in B$. Observe that $\lim_{n\to\infty} f_{n,k_n}(x) = f(x)$ for each $x \in B \cap \liminf_{n\to\infty} E_n$ and that $\lambda\left(B \cap \liminf_{n\to\infty} E_n\right) = \lambda\left(B \cap \bigcup_{m=1}^{\infty} \bigcap_{n=m}^{\infty} E_n\right) = 1$. It means that $f \in \overline{A}$ and finally $\overline{A} = \overline{\overline{A}}$. □

In Chap. 2 we shall return to this topic in more general setting (see Theorem 2.14).

Remark 1.16 ([BBT97], p. 427) No metric can describe pointwise convergence of functions defined on $[0, 1]$. Pointwise convergence can be considered as $\mathcal{I}$-a.e. convergence when a σ-ideal $\mathcal{I} = \{\varnothing\}$.

Definition 1.17 ([Wag81]) We shall say that a pair $(\mathcal{S}, \mathcal{I})$ fulfills the condition (E) if for each set $D \subset \mathcal{S} \setminus \mathcal{I}$ and for each double sequence $\{B_{j,n}\}_{j,n\in\mathbb{N}}$ of sets belonging to $\mathcal{S}$ and satisfying the following conditions: $B_{j,n} \subset B_{j,n+1}$ for each $j, n \in \mathbb{N}$, $\bigcup_{np=1}^{\infty} B_{j,n} = D$ for each $j \in \mathbb{N}$ there exists an increasing sequence $\{j_p\}_{p\in\mathbb{N}}$ of positive integers and a sequence $\{n_p\}_{p\in\mathbb{N}}$ of positive integers such that $\bigcap_{p=1}^{\infty} B_{j_p,n_p} \notin \mathcal{I}$.

Definition 1.18 ([Wag81]) We shall say that a pair $(\mathcal{S}, \mathcal{I})$ fulfills the condition (E') if for each set $D \subset \mathcal{S} \setminus \mathcal{I}$ and for each double sequence $\{B_{j,n}\}_{j,n\in\mathbb{N}}$ of sets belonging to $\mathcal{S}$ and satisfying the following conditions: $B_{j,n} \subset B_{j,n+1}$ for each $j, n \in \mathbb{N}$, $\bigcup_{n=1}^{\infty} B_{j,n} = D$ for each $j \in \mathbb{N}$ there exists a sequence $\{n_j\}_{j\in\mathbb{N}}$ of positive integers such that $\bigcap_{j=1}^{\infty} B_{j,n_j} \notin \mathcal{I}$.

Theorem 1.19 *(E) is equivalent to (E').*

Proof Obviously (E') implies (E). To prove that (E) implies (E') put $\hat{B}_{j,n} = \bigcap_{k=1}^{j} B_{k,n}$ for each $j, n \in \mathbb{N}$. Then $\hat{B}_{j,n} \in \mathcal{S}$, $\hat{B}_{j,n} \subset \hat{B}_{j,n+1}$ for each $j, n \in \mathbb{N}$ and $\bigcap_{n=1}^{\infty} \hat{B}_{j,n} = D$ for each $j \in \mathbb{N}$, moreover $\hat{B}_{j+1,n} \subset \hat{B}_{j,n}$ for $j, n \in \mathbb{N}$. By virtue of (E) there exists an increasing sequence $\{j_p\}_{p\in\mathbb{N}}$ of positive integers and a sequence $\{n_p\}_{p\in\mathbb{N}}$ of positive integers such that $\bigcap_{p=1}^{\infty} \hat{B}_{j_p,n_p} \notin \mathcal{I}$. Put $n(j) = n_p$ for $j_{p-1} < j \le j_p$ (assume that $j_0 = 0$). Then $\bigcap_{j=1}^{\infty} B_{j,n(j)} \supset \bigcap_{p=1}^{\infty} \hat{B}_{j_p,n_p}$, so $\bigcap_{j=1}^{\infty} B_{j,n(j)} \notin \mathcal{I}$. □

Theorem 1.20 *([Wag81]) $(\mathcal{L}, \mathcal{N})$ fulfills the condition (E), $(\mathcal{B}, \mathcal{K})$ does not fulfill (E).*

Proof We shall prove that $(\mathcal{L}, \mathcal{N})$ fulfills (E'). Suppose that a double sequence $\{B_{j,n}\}_{j,n\in\mathbb{N}}$ of measurable subsets of $[0, 1]$ fulfills both conditions and $\lambda(D) > 0$. Put $j_p = p$. Let $\varepsilon > 0$. For each $p \in \mathbb{N}$ there exists $n_p \in \mathbb{N}$ such that $\lambda(B_{p,n_p}) > \lambda(D) - \varepsilon \cdot 2^{-p}$. Then

$$\lambda\left(D \setminus \bigcap_{p=1}^{\infty} B_{p,n_p}\right) = \lambda\left(\bigcup_{p=1}^{\infty} (D \setminus B_{p,n_p})\right) \le \sum_{p=1}^{\infty} \lambda(D \setminus B_{p,n_p}) < \sum_{p=1}^{\infty} \varepsilon \cdot 2^{-p} = \varepsilon,$$

so $\lambda\left(\bigcap_{p=1}^{\infty} B_{p,n_p}\right) > \lambda(D) - \varepsilon > 0$ (in fact we proved that $\lambda\left(\bigcap_{p=1}^{\infty} B_{p,n_p}\right)$ can be made arbitrarily close to $\lambda(D)$.

Put $D = [0, 1]$ and

$$B_{j,n} = \left([0, 1] \setminus \bigcup_{i=1}^{j-1} \left(\frac{i}{j} - \frac{1}{2^n}, \frac{i}{j} + \frac{1}{2^n}\right)\right) \cup \left\{\frac{1}{j}, \frac{2}{j}, \ldots, \frac{j-1}{j}\right\}$$

for $j, n \in \mathbb{N}$. All sets $B_{j,n}$ have the Baire property and a double sequence $\{B_{j,n}\}_{j,n\in\mathbb{N}}$ fulfills both conditions and obviously D is not of the first category. Simultaneously for each increasing sequence $\{j_p\}_{p\in\mathbb{N}}$ of positive integers and a sequence $\{n_p\}_{p\in\mathbb{N}}$ of positive integers the set $\bigcap_{p=1}^{\infty} B_{j_p,n_p}$ is nowhere dense, because $D \setminus \bigcap_{p=1}^{\infty} B_{j_p,n_p}$ is open and dense in $[0, 1]$. □

Definition 1.21 We shall say that the pair $(\mathcal{S}, \mathcal{I})$ fulfills the countable chain condition (ccc) if each disjoint family sets from $\mathcal{S} \setminus \mathcal{I}$ is at most denumerable.

Theorem 1.22 *([W-BW81]) Suppose that the pair $(\mathcal{S}, \mathcal{I})$ fulfills (ccc). Then the convergence $\mathcal{I}$-a.e. of a sequence $\{f_n\}_{n\in\mathbb{N}}$ of real-valued $\mathcal{S}$-measurable functions defined on X to f implies the existence of a sequence $\{A_i\}_{i\in\mathbb{N}}$ of sets belonging to $\mathcal{S}$ such that $X \setminus \bigcup_{i=1}^{\infty} A_i \in \mathcal{I}$ and $\{f_n\}_{n\in\mathbb{N}}$ converges to f uniformly on each A_i if and only if the pair $(\mathcal{S}, \mathcal{I})$ fulfills the condition (E).*

Proof Suppose that the pair $(\mathcal{S}, \mathcal{I})$ fulfills (E) and that $\{f_n\}_{n\in\mathbb{N}}$ converges $\mathcal{I}$-a.e. to f. Put $E_{j,n} = \{x \in X \setminus A \colon |f_k(x) - f(x)| < \frac{1}{j}$ for each $k \geq n\}$, where $A \in \mathcal{I}$ is the set consisting of all points at which $\{f_n\}_{n\in\mathbb{N}}$ does not converge to f. Then $E_{j,n} \subset E_{j,n+1}$ for each $j, n \in \mathbb{N}$ and $\bigcup_{n=1}^{\infty} E_{j,n} = X \setminus A$ for each $j \in \mathbb{N}$. Obviously $E_{j,n} \in \mathcal{S}$ for each j, n. From the assumption it follows that there exists an increasing sequence $\{j_p\}_{p\in\mathbb{N}}$ of positive integers and a sequence $\{n_p\}_{p\in\mathbb{N}}$ of positive integers such that $\bigcap_{p=1}^{\infty} E_{j_p,n_p} = A_1 \in \mathcal{S} \setminus \mathcal{I}$. It is not difficult to observe that $\{f_n\}_{n\in\mathbb{N}}$ converges to f uniformly on A_1.

Suppose that for all ordinal numbers $\alpha < \eta$, where $\eta < \omega_1$, we have found sets $A_\alpha \in \mathcal{S} \setminus \mathcal{I}$ such that $\{f_n\}_{n\in\mathbb{N}}$ converges to f uniformly on each A_α. If $D_\eta = X \setminus \left(A \cup \bigcup_{\alpha<\eta} A_\alpha\right) \in \mathcal{I}$, then the proof is finished. Suppose that $D_\eta \notin \mathcal{I}$. Observe that $D_\eta \in \mathcal{S}$. Put $E^\eta_{j,n} = \{x \in D_\eta \colon |f_k(x) - f(x)| < \frac{1}{j}$ for each $k \geq n\}$. Then $E^\eta_{j,n} \subset E^\eta_{j,n+1}$ for each $j, n \in \mathbb{N}$ and $\bigcup_{n=1}^{\infty} E^\eta_{j,n} = D_\eta$ for each $j \in \mathbb{N}$ and also $E^\eta_{j,n} \in \mathcal{S}$ for each $j, n \in \mathbb{N}$. Again we can find an increasing sequence $\{j^\eta_p\}_{p\in\mathbb{N}}$ of positive integers and a sequence $\{n^\eta_p\}_{p\in\mathbb{N}}$ of positive integers such that $\bigcap_{p=1}^{\infty} E^\eta_{j_p,n_p} = A_\eta \in \mathcal{S} \setminus \mathcal{I}$. Then the sequence $\{f_n\}_{n\in\mathbb{N}}$ converges to f uniformly on A_η. Since the pair $(\mathcal{S}, \mathcal{I})$ fulfills (ccc), there exists an ordinal number $\beta < \omega_1$ such that $X \setminus \left(A \cup \bigcup_{\alpha<\beta} A_\alpha\right) \in \mathcal{I}$, so also $X \setminus \bigcup_{\alpha<\beta} A_\alpha \in \mathcal{I}$ and $\{f_n\}_{n\in\mathbb{N}}$ converges to f uniformly on each A_α, $\alpha < \beta$ and the family $\{A_\alpha\}_{\alpha<\beta}$ is at most denumerable.

Suppose now that the pair $(\mathcal{S}, \mathcal{I})$ does not fulfill the condition (E), so there exist a set $D \in \mathcal{S} \setminus \mathcal{I}$ and a double sequence $\{B_{j,n}\}_{j,n\in\mathbb{N}}$ of sets belonging to $\mathcal{S}$ such that $B_{j,n} \subset B_{j,n+1}$ for $j, n \in \mathbb{N}$, $\bigcup_{n=1}^{\infty} B_{j,n} = D$ for $j \in \mathbb{N}$ and for arbitrary increasing sequence $\{j_p\}_{p\in\mathbb{N}}$ of positive integers and arbitrary sequence $\{n_p\}_{p\in\mathbb{N}}$ of positive integers we have $\bigcap_{p=1}^{\infty} B_{j_p,n_p} \in \mathcal{I}$. Put $\hat{B}_{j,n} = \bigcap_{k=1}^{j} B_{k,n}$ for $j, n \in \mathbb{N}$. Observe that $\hat{B}_{j,n} \in \mathcal{S}$, $\hat{B}_{j,n} \subset \hat{B}_{j,n+1}$ for $j, n \in \mathbb{N}$ and $\bigcup_{n=1}^{\infty} \hat{B}_{j,n} = D$ for $j \in \mathbb{N}$. Moreover $\hat{B}_{j+1,n} \subset \hat{B}_{j,n}$ and $\hat{B}_{j,n} \subset B_{j,n}$ for $j, n \in \mathbb{N}$, so $\bigcap_{p=1}^{\infty} \hat{B}_{j_p,n_p} \in \mathcal{I}$ for arbitrary $\{j_p\}_{p\in\mathbb{N}}$ and $\{n_p\}_{p\in\mathbb{N}}$.

Observe that $\bigcup_{j=1}^{\infty} \left(\hat{B}_{j,n} \setminus \hat{B}_{j+1,n}\right) = \hat{B}_{1,n} \setminus \bigcap_{j=1}^{\infty} \hat{B}_{j,n}$ for each $n \in \mathbb{N}$. Indeed, if $x \in \bigcup_{j=1}^{\infty} \left(\hat{B}_{j,n} \setminus \hat{B}_{j+1,n}\right)$, then there exists $j_0 \in \mathbb{N}$ such that $x \in B_{j_0,n}$ and $x \notin \hat{B}_{j_0+1,n}$, so $x \notin \bigcap_{j=1}^{\infty} \hat{B}_{j,n}$ and finally $x \in \hat{B}_{1,n} \setminus \bigcap_{j=1}^{\infty} \hat{B}_{j,n}$. If $x \in \hat{B}_{1,n} \setminus \bigcap_{j=1}^{\infty} \hat{B}_{j,n}$, then there exists $j \in \mathbb{N}$ such that $x \notin \hat{B}_{j,n}$. Let j_0 be the smallest positive integer having this property. Obviously $j_0 \geq 2$, because $x \in \hat{B}_{1,n}$. Then $x \in \hat{B}_{j-1,n} \setminus \hat{B}_{j_0,n}$ and finally $x \in \bigcup_{j=1}^{\infty} \left(\hat{B}_{j,n} \setminus \hat{B}_{j+1,n}\right)$. Further $\bigcap_{j=1}^{\infty} \hat{B}_{j,n} \in \mathcal{I}$ and $D \setminus \bigcap_{j=1}^{\infty} \hat{B}_{j,n} \notin I$ from our assumption.

For every $n \in N$ put

$$f_n(x) = \begin{cases} \frac{1}{j} & \text{for } x \in \hat{B}_{j,n} \setminus \hat{B}_{j+1,n},\ j \in \mathbb{N} \\ 2 & \text{for } x \in D \setminus \hat{B}_{j,n} \\ 0 & \text{for } x \in (X \setminus D) \cup \bigcap_{j=1}^{\infty} \hat{B}_{j,n} \end{cases}.$$

If $x \in X \setminus D$, then obviously $f_n(x) \underset{n\to\infty}{\longrightarrow} 0$. For each $x \in D$ and $j \in \mathbb{N}$ there exists n_0 (depending on x and j) such that $x \in \hat{B}_{j,n}$ for $n \geq n_0$. Then from the definition it follows that $0 \leq f_n(x) \leq \frac{1}{j}$ and $f_n(x) \underset{n\to\infty}{\longrightarrow} 0$.

Observe also that $\{x \in X : 0 \leq f_n(x) \leq \frac{1}{j}\} = (X \setminus D) \cup B_{j,n}$ and, consequently,

$$\{x \in X : 0 \leq f_k(x) \leq \frac{1}{j} \text{ for each } k \geq n\} = (X \setminus D) \cup \bigcap_{k=n}^{\infty} \hat{B}_{j,k} = (X \setminus D) \cup \hat{B}_{j,n}.$$

Suppose now that $A \in \mathcal{S}$, $A \subset D$ is a set such that $\{f_n\}_{n\in\mathbb{N}}$ converges uniformly on A to zero. Then there exists a sequence $\{n_j\}_{j\in\mathbb{N}}$ of positive integers such that $A \subset \bigcap_{j=1}^{\infty} \hat{B}_{j,n_j}$. Hence $A \in \mathcal{I}$, and it is not possible to find a countable family $\{A_i\}_{i\in\mathbb{N}}$ of sets belonging to $\mathcal{S}$ and such that $D \setminus \bigcup_{i=1}^{\infty} A_i \in \mathcal{I}$ and $\{f_n\}_{n\in\mathbb{N}}$ converges uniformly to zero on each A_i, because $D \notin \mathcal{I}$. □

To simplify the denotation we shall say that a sequence $\{f_n\}_{n\in\mathbb{N}}$ of real-valued $\mathcal{S}$-measurable functions defined on X converges to f in the sense of Egorov if and only if there exists a sequence $\{A_i\}_{i\in\mathbb{N}}$ of sets belonging to $\mathcal{S}$ such that $X \setminus \bigcup_{i=1}^{\infty} A_i \in \mathcal{I}$ and $\{f_n\}_{n\in\mathbb{N}}$ converges to f uniformly on each A_i. Observe that without any assumptions concerning the pair $(\mathcal{S}, \mathcal{I})$ we have the following

Theorem 1.23 *If a sequence $\{f_n\}_{n\in\mathbb{N}}$ of real-valued $\mathcal{S}$-measurable functions defined on X converges to f in the sense of Egorov, then $\{f_n\}_{n\in\mathbb{N}}$ converges $\mathcal{I}$-a.e. to f.*

Definition 1.24 ([Tay60]) We shall say that a sequence $\{f_n\}_{n\in\mathbb{N}}$ of real-valued $\mathcal{S}$-measurable functions defined on X converges to f in the sense of Taylor if there exists a non-decreasing sequence $\{t_n\}_{n\in\mathbb{N}}$ tending to infinity such that the sequence $\{t_n \cdot (f_n - f)\}_{n\in\mathbb{N}}$ converges $\mathcal{I}$-a.e. to zero (compare [Vul76]).

Definition 1.25 ([Vul76]) We shall say that a sequence $\{f_n\}_{n\in\mathbb{N}}$ of real-valued $\mathcal{S}$-measurable functions defined on X converges to f with the convergence regulator if there exist a non-negative, a.e. finite $\mathcal{S}$-measurable function g and a sequence $\{\alpha_n\}_{n\in\mathbb{N}}$ of positive numbers convergent to zero such that

$$|f_n(x) - f(x)| \leq \alpha_n \cdot g(x)$$

for each $n \in \mathbb{N}$ and $x \in X$.

Definition 1.26 ([Yon75]) We shall say that a sequence $\{f_n\}_{n\in\mathbb{N}}$ of real-valued $\mathcal{S}$-measurable functions defined on X converges to f in the sense of Yoneda if there exists a non-negative $\mathcal{S}$-measurable function d defined on X such that for each $\varepsilon > 0$ there exists $n(\varepsilon) \in \mathbb{N}$ such that for each $n \geq n(\varepsilon)$ and for each $x \in X$ we have $|f_n(x) - f(x)| < \varepsilon \cdot d(x)$.

Theorem 1.27 *The following are equivalent:*

(a) $\{f_n\}_{n\in\mathbb{N}}$ *converges to* f *in the sense of Egorov,*
(b) $\{f_n\}_{n\in\mathbb{N}}$ *converges to* f *in the sense of Taylor,*
(c) $\{f_n\}_{n\in\mathbb{N}}$ *converges to* f *with some convergence regulator,*
(a) $\{f_n\}_{n\in\mathbb{N}}$ *converges to* f *in the sense of Yoneda.*

Proof (a)$\Rightarrow$(b) Let $\{A_m\}_{m\in\mathbb{N}}$ be a sequence of sets from $\mathcal{S}$ such that $X \setminus \bigcup_{m=1}^{\infty} A_m \in \mathcal{I}$ and the convergence is uniform on each A_m. For each $m \in \mathbb{N}$ choose an increasing sequence $\{n_{m,r}\}_{r\in\mathbb{N}}$ of positive integers such that $|f_n(x) - f(x)| < \frac{1}{r+1}$ for each $x \in A_m$ and each $n \geq n_{m,r}$. Let $\{n_r\}_{r\in\mathbb{N}}$ be an increasing sequence of positive integers such that $\lim_{r\to\infty} \frac{n_r}{n_{m,r}} = \infty$ for each $m \in \mathbb{N}$. Put

$$t_n = \begin{cases} 1 & \text{for } 1 \leq n < n_1 \\ \sqrt{r} & \text{for } n_{r-1} \leq n < n_r, \quad \text{for } r \geq 2. \end{cases}$$

If $x \in \bigcup_{m=1}^{\infty} A_m$, then $x \in A_m$ for some $m \in \mathbb{N}$. Then there exists $r_0 \in \mathbb{N}$ such that $n_r \geq n_{m,r}$ for $r \geq r_0$. If $n \geq n_{r_0}$, then there exists $r \geq r_0$ such that $n_r \leq n < n_{r+1}$. Hence for such n we have $|f_n(x) - f(x)| < \frac{1}{r+1} = \frac{1}{t_n^2}$, so $t_n \cdot |f_n(x) - f(x)| < \frac{1}{t_n}$. Finally $\{t_n \cdot |f_n - f|\}_{n\in\mathbb{N}}$ converges to zero on $\bigcup_{m=1}^{\infty} A_m$, which means $\mathcal{I}$-a.e.

(b)$\Rightarrow$(c) Put $g(x) = \sup_n t_n \cdot |f_n(x) - f(x)|$ for $x \in X$ and $\alpha_n = \frac{1}{t_n}$. Obviously g is $\mathcal{S}$-measurable and by virtue of (b) g is $\mathcal{I}$-a.e. finite. The inequality $|f_n(x) - f(x)| \leq \frac{1}{t_n} \cdot g(x)$ is obvious.

(c)$\Rightarrow$(d) Put $d(x) = g(x)$. Take $\varepsilon > 0$. There exists n_ε such that $\alpha_n = \frac{1}{t_n} < \varepsilon$ for $n \geq n_\varepsilon$. Then obviously $|f_n(x) - f(x)| < \varepsilon \cdot d(x)$ for each $n \geq n_\varepsilon$ and each $x \in X$.

(d) $\Rightarrow$ (a) Put $A_m = \{x \in X\colon d(x) \leq m\}$. Then $X \setminus \bigcup_{m=1}^{\infty} A_m \in \mathcal{I}$ and $\{f_n\}_{n\in\mathbb{N}}$ converges uniformly to f on each A_m. □

Chapter 2
Convergence in Measure and in Category

The notion of the convergence in measure of sequences of real-valued measurable functions has been introduced and studied by F. Riesz. In his definition the essential role is played by the measure of sets of the form $\{x\colon |f_n(x) - f(x)| > \varepsilon\}$. From Theorems 2.5 and 2.6 it follows that all what is necessary is σ-algebra of sets and σ-ideal $\mathcal{I} \subset \mathcal{S}$. Below we shall present the convergence with respect to an abstract σ-ideal introduced in [Wag81] and study the most interesting for us examples when $(\mathcal{S}, \mathcal{I}) = (\mathcal{L}, \mathcal{N})$ or $(\mathcal{S}, \mathcal{I}) = (\mathcal{B}, \mathcal{K})$.

Definition 2.1 We shall say that a sequence $\{f_n\}_{n\in\mathbb{N}}$ of real-valued $\mathcal{S}$-measurable functions defined on $[0, 1]$ converges in measure to a function f if

$$\lim_{n\to\infty} \lambda\left(\{x \in [0, 1]\colon |f_n(x) - f(x)| > \varepsilon\}\right) = 0$$

for each $\varepsilon > 0$.

Equivalently: for each $\varepsilon > 0$ and $\eta > 0$ there exists $n_0 \in \mathbb{N}$ such that

$$\lambda\left(\{x \in [0, 1]\colon |f_n(x) - f(x)| > \varepsilon\}\right) < \eta$$

for $n \geq n_0$.

Proposition 2.2 *If a sequence* $\{f_n\}_{n\in\mathbb{N}}$ *of real-valued* $\mathcal{S}$*-measurable functions defined on* $[0, 1]$ *converges in measure to measurable functions* f *and* g*, then* $f = g$ *almost everywhere.*

Proof Take $\varepsilon > 0$. We have

$$\{x \in [0, 1]\colon |f(x) - g(x)| > \varepsilon\} \subset \left\{x \in [0, 1]\colon |f_n(x) - f(x)| > \frac{\varepsilon}{2}\right\} \cup \left\{x \in [0, 1]\colon |f_n(x) - g(x)| > \frac{\varepsilon}{2}\right\}$$

W. Wilczyński, *Convergence in Measure and in Category*, Lecture Notes in Mathematics 2394, https://doi.org/10.1007/978-3-032-23143-7_2

for each $n \in \mathbb{N}$. Hence

$$\lambda(\{x \in [0,1] \colon |f(x) - g(x)| > \varepsilon\}) \leq \lambda\left(\left\{x \in [0,1] \colon |f_n(x) - f(x)| > \frac{\varepsilon}{2}\right\}\right) + \lambda\left(\left\{x \in [0,1] \colon |f_n(x) - g(x)| > \frac{\varepsilon}{2}\right\}\right)$$

for each $n \in \mathbb{N}$ and $\lambda(\{x \in [0,1] \colon |f(x) - g(x)| > \varepsilon\}) = 0$. From the arbitrariness of $\varepsilon > 0$ it follows immediately that $f = g$ a.e. □

Theorem 2.3 *If a sequence* $\{f_n\}_{n\in\mathbb{N}}$ *of real-valued measurable functions defined on* $[0,1]$ *converges to* f *almost everywhere, then* $\{f_n\}_{n\in\mathbb{N}}$ *converges to* f *in measure.*

Proof Take $\varepsilon > 0$. Put $E_n = \{x \in [0,1] \colon |f_n(x) - f(x)| > \varepsilon\}$ for each $n \in \mathbb{N}$ and $E_0 = \{x \in [0,1] \colon \lim_{n\to\infty} f_n(x) = f(x)\}$. Then $E_0 \cap \limsup_{n\to\infty} E_n = \varnothing$. Since $\lambda(E_0) = 1$, we have $\lambda(\limsup_{n\to\infty} E_n) = 0$. Hence

$$0 = \lambda(\limsup_{n\to\infty} E_n) = \lambda\left(\bigcap_{n=1}^{\infty}\bigcup_{k=n}^{\infty} E_k\right) = \lim_{n\to\infty} \lambda\left(\bigcup_{k=n}^{\infty} E_k\right),$$

so also $\lim_{n\to\infty} \lambda(E_n) = 0$. □

Remark 2.4 If the domain of f_n's is not of finite measure, the thesis does not hold. For example, the sequence $\{\chi_{[n,\infty)}\}_{n\in\mathbb{N}}$ of functions defined on $[0,\infty)$ converges a.e. but not in measure to zero.

Theorem 2.5 (Riesz) *If a sequence* $\{f_n\}_{n\in\mathbb{N}}$ *of real-valued measurable functions defined on* $[0,1]$ *converges in measure to a function* f*, then there exists an increasing sequence* $\{n_m\}_{m\in\mathbb{N}}$ *of positive integers such that* $\{f_{n_m}\}_{m\in\mathbb{N}}$ *converges to* f *almost everywhere.*

Proof For each $m \in \mathbb{N}$ there exists $n_m \in \mathbb{N}$ such that

$$\lambda\left(\left\{x \in [0,1] \colon |f_{n_m}(x) - f(x)| > \frac{1}{2^m}\right\}\right) < \frac{1}{2^m}.$$

Obviously we can suppose that $n_{m+1} > n_m$. Denote $E_m = \{x \in [0,1] \colon |f_{n_m}(x) - f(x)| > \frac{1}{2^m}\}$. Then

$$E = \limsup_{m\to\infty} E_m = \bigcap_{m=1}^{\infty}\bigcup_{k=m}^{\infty} E_k$$

is a null set because

$$\lambda(E) = \lim_{m\to\infty} \lambda\left(\bigcup_{k=m}^{\infty} E_k\right)$$

and

$$\lambda\left(\bigcup_{k=m}^{\infty} E_k\right) < \frac{1}{2^{m-1}}.$$

Hence $\lambda([0,1] \setminus E) = 1$. If

$$x \in [0,1] \setminus E = \liminf_{m\to\infty}([0,1] \setminus E_m) = \bigcap_{m=1}^{\infty}\bigcup_{k=m}^{\infty}([0,1] \setminus E_k),$$

then obviously $\lim_{m\to\infty} f_{n_m}(x) = f(x)$. □

Theorem 2.6 *A sequence $\{f_n\}_{n\in\mathbb{N}}$ of real-valued measurable functions defined on $[0,1]$ converges in measure to a function f if and only if for each subsequence $\{f_{n_m}\}_{m\in\mathbb{N}}$ there exists a subsequence $\{f_{n_{m_p}}\}_{p\in\mathbb{N}}$ which converges a.e. to f.*

Proof Necessity follows from theorem of Riesz and from the obvious fact that convergence in measure is preserved by subsequences.

To prove the sufficiency suppose that $\{f_n\}_{n\in\mathbb{N}}$ does not converge in measure to f. Then from the definition it follows immediately that there exist a pair of positive numbers ε and δ and an increasing sequence $\{n_m\}_{m\in\mathbb{N}}$ of positive integers such that

$$\lambda\left(\{x \in [0,1] \colon |f_{n_m}(x) - f(x)| > \varepsilon\}\right) > \delta$$

for each $m \in \mathbb{N}$. Suppose that $\{f_{n_{m_p}}\}_{p\in\mathbb{N}}$ converges a.e. to f. Then

$$\lim_{p\to\infty} \lambda\left(\{x \in [0,1] \colon |f_{n_{m_p}}(x) - f(x)| > \varepsilon\}\right) = 0,$$

a contradiction. □

Remark 2.7 There exist sequences of real-valued measurable functions convergent in measure but not almost everywhere.

The commonly presented example looks as follows: each positive integer can be uniquely represented in the form $n = 2^k + i$, where $k \in \mathbb{N} \cup \{0\}$ and $0 \le i \le 2^k$. Put $f_n = \chi_{\left[\frac{i}{2^k}, \frac{i+1}{2^k}\right]}$ for $n \in \mathbb{N}$, where k and i are from the above representation. Then $\liminf_{n\to\infty} f_n(x) = 0$ and $\limsup_{n\to\infty} f_n(x) = 1$ for each $x \in [0,1]$, so $\{f_n\}_{n\in\mathbb{N}}$ diverges everywhere. Simultaneously $\lim_{n\to\infty} \lambda(\{x \in [0,1] \colon f_n(x) \ne 0\}) = 0$, so $\{f_n\}_{n\in\mathbb{N}}$ converges to 0 in measure.

This example was constructed in the proof of Theorem 1.13.

For every measurable real-valued function f defined on $[0,1]$ put

$$\| f \| = \inf\{\varepsilon > 0 \colon \lambda(\{x \colon |f(x)| > \varepsilon\}) < \varepsilon\}.$$

Observe that the inequality $\lambda(\{x \colon |f(x)| > \varepsilon\}) < \varepsilon$ is fulfilled if ε is sufficiently large and if it is satisfied for some $\varepsilon > 0$, then is also satisfied for each larger value. Hence $0 \le \| f \| < \infty$ for each f.

Proposition 2.8

(1) $\| f \| = 0$ *if and only if* $f = 0$ *a.e.*
(2) $\| f + g \| \leq \| f \| + \| g \|$ *for each* f, g.
(3) $\lambda(\{x\colon |f(x)| > \| f \|\}) \leq \| f \|$.
(4) $\| cf \| \leq \max(\| f \|, c \| f \|)$.

Proof (1) If $\| f \| = 0$, then $\| f \| < \frac{1}{n}$ for each $n \in \mathbb{N}$. So for each $n \in \mathbb{N}$ $\lambda\left(\{x\colon |f(x)| > \frac{1}{n}\}\right) < \frac{1}{n}$ and since $\{x\colon |f(x) > 0\} = \bigcup_n \{x\colon |f(x)| > \frac{1}{n}\}$ we obtain $\lambda(\{x\colon |f(x)| > 0\}) = 0$. Conversely, if $f = 0$ a.e., then for each $\varepsilon > 0$ we have $\lambda(\{x\colon |f(x)| > \varepsilon\}) < \varepsilon$, so $\| f \| = 0$.

(2) Take $\varepsilon_f > \| f \|$ and $\varepsilon_g > \| g \|$. Then

$$\lambda\left(\{x\colon |f(x) + g(x)| > \varepsilon_f + \varepsilon_g\}\right) \leq \lambda\left(\{x\colon |f(x)| > \varepsilon_f\}\right) + \lambda\left(\{x\colon |g(x)| > \varepsilon_g\}\right) < \varepsilon_f + \varepsilon_g$$

and from arbitrariness of ε_f and ε_g it follows that $\| f + g \| \leq \| f \| + \| g \|$.

(3) Let $\{\varepsilon_n\}_{n\in\mathbb{N}}$ be a decreasing sequence convergent to $\| f \|$. Then we have

$$\{x\colon |f(x)| > \| f \|\} = \bigcup_{n=1}^{\infty} \{x\colon |f(x)| > \varepsilon_n\}$$

and $\lambda(\{x\colon |f(x)| > \varepsilon_n\}) < \varepsilon_n$ for each $n \in \mathbb{N}$. Hence

$$\lambda(\{x\colon |f(x)| > \| f \|\}) = \lim_{n\to\infty} \lambda(\{x\colon |f(x)| > \varepsilon_n\}) \leq \lim_{n\to\infty} \varepsilon_n = \| f \| .$$

(4) If $c \leq 1$, then $\{x\colon |cf(x)| > \varepsilon\} \subset \{x\colon |f(x)| > \varepsilon\}$, so if $\lambda(\{x\colon |f(x)| > \varepsilon\}) < \varepsilon$, then $\lambda(\{x\colon |cf(x)| > \varepsilon\}) < \varepsilon$ and finally $\| cf \| \leq \| f \|$.

Suppose now that $c > 1$. Obviously $\{x\colon |cf(x)| > c \| f \|\} = \{x\colon |f(x)| \geq \|f \|\}$, so from (3) $\lambda(\{x\colon |cf(x)| > c \| f \|\}) = \lambda(\{x\colon |f(x)| > \| f \|\}) \leq \| f \| < c \| f \|$. Hence $\| cf \| = \inf\{\varepsilon\colon \lambda(\{x\colon |cf(x)| > \varepsilon\}) \leq c \| f \|$. □

Remark 2.9 The functional $\| \cdot \|$ is not a norm, because it is not homogeneous. For example if $f(x) = x$, then $\| f \| = \frac{1}{2}$ and if $g(x) = 2f(x) = 2x$, then $\| g \| = \frac{2}{3} \neq 2 \| f \|$.

From the above proposition it follows that $\rho(f, g) = \| f - g \|$ is a pseudometric.

Theorem 2.10 *A sequence* $\{f_n\}_{n\in\mathbb{N}}$ *of real-valued measurable functions defined on* $[0, 1]$ *converges in measure to* f *if and only if* $\lim_{n\to\infty} \| f_n - f \| = 0$.

Proof Suppose that $f_n \to f$ in measure. Then for each $\varepsilon > 0$

$$\lambda(\{x\colon |f_n(x) - f(x)| > \varepsilon\}) \underset{n\to\infty}{\longrightarrow} 0,$$

so there exists $n_\varepsilon \in \mathbb{N}$ such that $\lambda(\{x\colon |f_n(x) - f(x)| > \varepsilon\}) < \varepsilon$ for each $n \geq n_\varepsilon$ and $\| f_n - f \| \leq \varepsilon$ for $n \geq n_\varepsilon$

Suppose now that $\lim_{n\to\infty} \| f_n - f \| = 0$. For each $\mu > 0$ there exists $n_\mu \in \mathbb{N}$ such that $\| f_n - f \| < \mu$ for $n \geq n_\mu$. Hence $\lambda(\{x : |f_n(x) - f(x)| > \mu\}) < \mu$ for such n. Take arbitrary $\varepsilon > 0$ and $\eta > 0$ and put $\mu = \min(\varepsilon, \eta)$. For $n \geq n_\mu$ we have $\lambda(\{x : |f_n(x) - f(x)| > \mu\}) < \mu$. Since

$$\{x : |f_n(x) - f(x)| > \varepsilon\} \subset \{x : |f_n(x) - f(x)| > \mu\}$$

we have

$$\lambda(\{x : |f_n(x) - f(x)| > \varepsilon\}) \leq \lambda(\{x : |f_n(x) - f(x)| > \mu\}) < \mu \leq \eta,$$

which means $f_n \to f$ in measure. □

Theorem 2.11 *A sequence $\{f_n\}_{n\in\mathbb{N}}$ of real-valued measurable functions defined on $[0, 1]$ converges in measure to f if and only if $\lim_{n\to\infty} \int_0^1 \frac{|f_n(x)-f(x)|}{1+|f_n(x)-f(x)|} dx = 0$.*

Proof Observe that a function $\varphi(y) = \frac{y}{1+y}$ is increasing in the half-line $[0, \infty)$ and $\varphi(y) < 1$ for each $y \in [0, \infty)$.

Suppose that $\lim_{n\to\infty} \int_0^1 \frac{|f_n(x)-f(x)|}{1+|f_n(x)-f(x)|} dx = 0$. Fix $\varepsilon > 0$. Let $A_n(\varepsilon) = \{x \in [0, 1] : |f_n(x) - f(x)| > \varepsilon\}$ for $\varepsilon > 0$. Then we have

$$\frac{\varepsilon}{1+\varepsilon} < \frac{|f_n(x) - f(x)|}{1 + |f_n(x) - f(x)|}$$

for $x \in A_n(\varepsilon)$. Hence

$$\lambda(A_n(\varepsilon)) \cdot \frac{\varepsilon}{1+\varepsilon} = \int_{A_n(\varepsilon)} \frac{\varepsilon}{1+\varepsilon} dx < \int_0^1 \frac{|f_n(x) - f(x)|}{1 + |f_n(x) - f(x)|} dx$$

so $\lim_{n\to\infty} \lambda(A_n(\varepsilon)) = 0$ which means that $\{f_n\}_{n\in\mathbb{N}}$ converges to f in measure.

Suppose now that $\{f_n\}_{n\in\mathbb{N}}$ converges to f in measure. Then $\lim_{n\to\infty} \lambda(A_n(\varepsilon)) = 0$ for each $\varepsilon > 0$. If $x \in A_n(\varepsilon)$, then obviously $\frac{|f_n(x)-f(x)|}{1+|f_n(x)-f(x)|} < 1$. If $x \in [0, 1] \setminus A_n(\varepsilon)$, then $\frac{|f_n(x)-f(x)|}{1+|f_n(x)-f(x)|} < \frac{\varepsilon}{1+\varepsilon} < \varepsilon$. Hence

$$\int_0^1 \frac{|f_n(x) - f(x)|}{1 + |f_n(x) - f(x)|} dx < \int_{A_n(\varepsilon)} dx + \int_{[0,1]\setminus A_n(\varepsilon)} \varepsilon dx < \lambda(A_n(\varepsilon)) + \varepsilon.$$

From the arbitrariness of ε and from the assumption it follows that

$$\lim_{n\to\infty} \int_0^1 \frac{|f_n(x) - f(x)|}{1 + |f_n(x) - f(x)|} dx = 0.$$

□

Put $d(f, g) = \int_0^1 \frac{|f(x)-g(x)|}{1+|f(x)-g(x)|}dx$ for real-valued measurable function f, g defined on $[0, 1]$. It is not difficult to check that d is a pseudometric. The above theorem says that a sequence $\{f_n\}_{n\in\mathbb{N}}$ converges to f in measure if and only if $\lim_{n\to\infty} d(f_n, f) = 0$.

If $\lambda(\{x \in [0, 1]\colon f_1(x) \neq f_2(x)\} = 0$, then we call f_1 and f_2 equivalent. The class of measurable functions equivalent to f will be denoted by $[f]$, and the family of equivalence classes will be designated by $\mathcal{M}$. We shall say that a sequence $\{[f_n]\}_{n\in\mathbb{N}}$ of elements of $\mathcal{M}$ converges to $[f]$ in measure if and only if $\{f_n\}_{n\in\mathbb{N}}$ converges in measure to f. If we put $\rho([f], [g]) = \rho(f, g)$ and $d([f], [g]) = d(f, g)$, we obtain two equivalent metrics describing the convergence in measure of elements of $\mathcal{M}$.

From Theorem 2.6 it follows that the convergence in measure can be described using only σ-ideal $\mathcal{N}$ of sets of measure zero. This enables to define the convergence with respect to σ-ideal in an abstract setting. Suppose now that $(X, \mathcal{S})$ is a measurable space equipped with σ-ideal $\mathcal{I} \subset \mathcal{S}$.

Definition 2.12 ([Wag81]) We shall say that the sequence $\{f_n\}_{n\in\mathbb{N}}$ of real-valued $\mathcal{S}$-measurable functions defined on X converges with respect to $\mathcal{I}$ to a real-valued $\mathcal{S}$-measurable function f defined on X if and only if for each subsequence $\{f_{n_m}\}_{m\in\mathbb{N}}$ of $\{f_n\}_{n\in\mathbb{N}}$ there exists a subsequence $\{f_{n_{m_p}}\}_{p\in\mathbb{N}}$ which converges to f $\mathcal{I}$-a.e. We shall use the denotation $f_n \xrightarrow[n\to\infty]{\mathcal{I}} f$ (or $\lim_{n\to\infty} f_n = f$ with respect to $\mathcal{I}$).

Remark 2.13 From this definition it follows immediately that the convergence $\mathcal{I}$-a.e. implies the convergence with respect to $\mathcal{I}$.

If $f = g$ $\mathcal{I}$-a.e., then we shall say these functions are equivalent ($f \sim g$). If $f_n \sim h_n$ for $n \in \mathbb{N}$, $f \sim g$ and $f_n \xrightarrow[n\to\infty]{\mathcal{I}} f$ (or $\lim_{n\to\infty} f_n = f$ $\mathcal{I}$-a.e.), then also $h_n \xrightarrow[n\to\infty]{\mathcal{I}} g$ (or $\lim_{n\to\infty} h_n = g$ $\mathcal{I}$-a.e.). So if $[f] = \{g\colon f \sim g\}$ denotes the equivalence class of a real-valued $\mathcal{S}$-measurable function defined on X in the family $\mathcal{M}$ of all real-valued $\mathcal{S}$-measurable functions defined on X, then we can write $[f_n] \xrightarrow[n\to\infty]{\mathcal{I}} [f]$ if and only if $f_n \xrightarrow[n\to\infty]{\mathcal{I}} f$. However, we shall formulate all statements in terms of functions rather than in terms of equivalence classes.

The following conditions are fulfilled:

(L1) If $f_n = f$ for each $n \in \mathbb{N}$, then $f_n \xrightarrow[n\to\infty]{\mathcal{I}} f$.

(L2) If $f_n \xrightarrow[n\to\infty]{\mathcal{I}} f$ and $\{f_{n_m}\}_{n\in\mathbb{N}}$ is a subsequence of $\{f_n\}_{n\in\mathbb{N}}$, then $f_{n_m} \xrightarrow[n\to\infty]{\mathcal{I}} f$.

(L3) If $\{f_n\}_{n\in\mathbb{N}}$ does not converge to f with respect to $\mathcal{I}$, then there exists a subsequence $\{f_{n_m}\}_{n\in\mathbb{N}}$ no subsequence of which converges to f with respect to $\mathcal{I}$.

So the set of $\mathcal{S}$-measurable real-valued functions defined on X equipped with the convergence with respect to $\mathcal{I}$ is an $\mathcal{L}^*$ space (see [Eng89]). One can define in such a space the closure operation assuming that f belongs to the closure of set $A \subset \mathcal{M}$ ($f \in \overline{A}$) if and only if there exists a sequence $\{f_n\}_{n\in\mathbb{N}}$ of functions from A such that $f_{n_m} \xrightarrow[n\to\infty]{\mathcal{I}} f$ (or, equivalently $\lim_{n\to\infty} f_n = f$ $\mathcal{I}$-a.e.). This operation has the following properties: $\overline{\varnothing} = \varnothing$, $A \subset \overline{A}$ and $\overline{A \cup B} = \overline{A} \cup \overline{B}$. The fourth axiom of Kuratowski $\overline{\overline{A}} = \overline{A}$ is fulfilled if and only if the following condition holds:

(L4) If $f_j \xrightarrow[j\to\infty]{\mathcal{I}} f$ and $f_{j,n} \xrightarrow[n\to\infty]{\mathcal{I}} f_j$ for each $j \in \mathbb{N}$, then there exist two sequences $\{j_p\}_{p\in\mathbb{N}}$ and $\{n_p\}_{p\in\mathbb{N}}$ of positive integers such that $f_{j_p,n_p} \xrightarrow[p\to\infty]{\mathcal{I}} f$.

The space $\mathcal{L}^*$ fulfilling (L4) can be equipped with the topology determined by the closure operation described above. This topology usually is called a Frechet topology.

Theorem 2.14 *Suppose that the pair $(\mathcal{S}, \mathcal{I})$ fulfills (ccc). Then the family of real-valued $\mathcal{S}$-measurable functions defined on X is equipped with the Frechet topology generated by the convergence with respect to $\mathcal{I}$ if and only if the pair $(\mathcal{S}, \mathcal{I})$ fulfills the condition (E) (or, equivalently (E')) (see Definitions 1.17 and 1.18).*

Proof Sufficiency. Suppose that $f_j \xrightarrow[j\to\infty]{\mathcal{I}} f$ and $f_{j,n} \xrightarrow[n\to\infty]{\mathcal{I}} f_j$ for each $j \in \mathbb{N}$. Choosing, if necessary, subsequences we can assume that all sequences above converge $\mathcal{I}$-a.e.. Let $A_0 = \{x \in X\colon f_j(x)$ does not converge to $f(x)\}$ and $A_j = \{x \in X\colon f_{j,n}(x)$ does not converge to $f_j(x)\}$. Put $A = \bigcup_{j=0}^{\infty} A_j$. Obviously $A \in \mathcal{I}$. We shall need the following lemma. □

Lemma 2.15 *Under the above assumptions $f_{j,n}$, f_j and f for each set $D \in \mathcal{S} \setminus \mathcal{I}$, $D \subset X \setminus A$ there exists a set $B \subset D$, $B \in \mathcal{S} \setminus \mathcal{I}$ and a sequence $\{n_j\}_{j\in\mathbb{N}}$ of positive integers such that for each sequence $\{n'_j\}_{j\in\mathbb{N}}$ of positive integers for which $n_j \le n'_j$ except for finite number of $j \in \mathbb{N}$ the sequence $\{f_{j,n_j}\}_{j\in\mathbb{N}}$ converges to f everywhere on B.*

Proof of the Lemma Put $B_{j,n} = \{x \in D\colon |f_{j,k}(x) - f_j(x)| < \frac{1}{j}$ for each $k \ge n\}$. Then $B_{j,n} \subset B_{j,n+1}$ for $j, n \in \mathbb{N}$ and $\bigcup_{n=1}^{\infty} B_{j,n} = D$ for each $j \in \mathbb{N}$. From the condition (E') it follows that there exists a sequence $\{n_j\}_{j\in\mathbb{N}}$ of positive integers such that $B = \bigcap_{j=1}^{\infty} B_{j,n_j} \notin \mathcal{I}$. If $n_j \le n'_j$ for almost all $j \in \mathbb{N}$ and $x \in B$, then $|f_{j,n'_j}(x) - f_j(x)| < \frac{1}{j}$ for almost all $j \in \mathbb{N}$. Also we have $\lim_{j\to\infty} f_j(x) = f(x)$, so $\lim_{j\to\infty} f_{j,n_j}(x) = f(x)$ for $x \in B$. □

Let us return to the proof of Theorem 2.14. Using the above lemma we shall show the existence of a sequence $\{n_j\}_{j\in\mathbb{N}}$ of positive integers for which $f_{j,n_j} \xrightarrow[j\to\infty]{} f$ $\mathcal{I}$-a.e. Take $D_0 = X \setminus A$. By virtue of Lemma 2.15 there exists a set $B_0 \subset D$, $B_0 \in \mathcal{S} \setminus \mathcal{I}$ and a sequence $\{n_j^0\}_{j\in\mathbb{N}}$ of positive integers such that if $n_j^0 \le n'_j$ for almost all $j \in \mathbb{N}$, then $\lim_{j\to\infty} f_{j,n_j^0}(x) = f(x)$ for each $x \in B_0$. Suppose that for each ordinal $\beta < \alpha$, where $\alpha < \omega_1$, we have found a set $B_\beta \subset D$, $B_\beta \in \mathcal{S} \setminus \mathcal{I}$ and a sequence $\{n_j^\beta\}_{j\in\mathbb{N}}$ of positive integers for which $B_{\beta_1} \cap B_{\beta_2} = \varnothing$ and $n_j^{\beta_1} \le n_j^{\beta_2}$ for almost all

$j \in \mathbb{N}$ when $\beta_1 < \beta_2 < \alpha$ and $\lim_{j\to\infty} f_{j,n'_j}(x) = f(x)$ for $x \in \bigcup_{\eta\le\beta} B_\eta$ if $n_j^\beta \le n'_j$ for almost all $j \in \mathbb{N}$ for each $\beta < \alpha$. Put $D_\alpha = X \setminus \left(A \cup \bigcup_{\beta<\alpha} B_\beta\right)$. If $D_\alpha \in \mathcal{I}$, then we stop. If not and if α has a predecessor β_0 then from Lemma 2.15 we conclude that there exist a set $B_\alpha \subset D_\alpha$, $B_\alpha \in \mathcal{S}\setminus\mathcal{I}$ and a sequence $\{n_j^\alpha\}_{j\in\mathbb{N}}$ such that $\lim_{j\to\infty} f_{j,n_j^\alpha}(x) = f(x)$ for $x \in B_\alpha$. Obviously we have $\lim_{j\to\infty} f_{j,n'_j}(x) = f(x)$ for $x \in \bigcup_{\beta\le\alpha} B_\beta$ and if $n_j^\alpha \le n'_j$ for almost all $j \in \mathbb{N}$. If $\alpha < \omega_1$ is a limit number, then there exists an increasing sequence $\{\beta_k\}_{k\in\mathbb{N}}$ such that $\alpha = \lim_{k\to\infty}\beta_k$. With each $k \in \mathbb{N}$ there is associated a sequence $\{n_j^{\beta_k}\}_{j\in\mathbb{N}}$ such that $n_j^{\beta_k} \le n_j^{\beta_{k+1}}$ for almost all $j \in \mathbb{N}$ for each $k \in \mathbb{N}$ and also $\lim_{j\to\infty} f_{j,n_j^{\beta_k}}(x) = f(x)$ for $x \in \bigcup_{\eta\le\beta_k} B_\eta$. Put $n_j^\alpha = \max\{n_j^{\beta_1}, \dots, n_j^{\beta_j}\}$. Then we have $n_j^{\beta_k} \le n_j^\alpha$ for almost all $j \in \mathbb{N}$ and each $k \in \mathbb{N}$, so $\lim_{j\to\infty} f_{j,n_j^\alpha}(x) = f(x)$ for $x \in \bigcup_{k\in\mathbb{N}}\bigcup_{\eta\le\beta_k} B_\eta = \bigcup_{\beta\le\alpha} B_\beta$. Since $(\mathcal{S},\mathcal{I})$ fulfills (ccc), there exists $\alpha < \omega_1$ for which $D_\alpha \in \mathcal{I}$ which means that $\{f_{j,n_j^\alpha}\}_{j\in\mathbb{N}}$ converges to f $\mathcal{I}$-a.e. on X.
Necessity. Suppose that (E) does not hold. There exists a set $D \in \mathcal{S}\setminus\mathcal{I}$ and a double sequence $\{B_{j,n}\}_{j,n\in\mathbb{N}}$ of sets from $\mathcal{S}$ such that $B_{j,n} \subset B_{j,n+1}$ for each $j, n \in \mathbb{N}$ and $\bigcup_{n=1}^\infty B_{j,n} = D$ for $j \in \mathbb{N}$ and for each increasing sequence $\{j_p\}_{p\in\mathbb{N}}$ of positive integers and for each sequence $\{n_p\}_{p\in\mathbb{N}}$ of positive integers $\bigcap_{p=k}^\infty B_{j_p,n_p} \in \mathcal{I}$. Observe that also $\bigcap_{p=k}^\infty B_{j_p,n_p} \in \mathcal{I}$ for each $k \in \mathbb{N}$. Put $f_{j,n}(x) = \chi_{D\setminus B_{j,n}}(x) + \frac{1}{j}$ for $j, n \in \mathbb{N}$. Then $\lim_{n\to\infty} f_{j,n}(x) = f_j(x) = \frac{1}{j}$ for each $j \in \mathbb{N}$ and $\lim_{j\to\infty} f_j = f = 0$ and all limits exist for each $x \in X$ (so all sequences are convergent with respect to $\mathcal{I}$). If $x \in \limsup_{p\to\infty}\left(D \setminus B_{j_p,n_p}\right)$, then $\lim_{p\to\infty} f_{j_p,n_p}(x) = 1$. By virtue of the above observation

$$\limsup_{p\to\infty}\left(D \setminus B_{j_p,n_p}\right) = \bigcap_{k=1}^\infty\bigcup_{p=k}^\infty\left(D \setminus B_{j_p,n_p}\right) = D \setminus \bigcup_{k=1}^\infty\bigcap_{p=k}^\infty\left(D \setminus B_{j_p,n_p}\right)$$

and $\bigcup_{k=1}^\infty\bigcap_{p=k}^\infty\left(D \setminus B_{j_p,n_p}\right) \in \mathcal{I}$, so $\{x \in X\colon \limsup_{p\to\infty} f_{j_p,n_p}(x) \ne f(x)\} \in \mathcal{S}\setminus\mathcal{I}$ and from the arbitrariness of $\{j_p\}_{p\in\mathbb{N}}$ and $\{n_p\}_{p\in\mathbb{N}}$ it follows that the condition $(L4)$ is not fulfilled.

If $X = [0, 1]$ and $(\mathcal{S},\mathcal{I}) = (\mathcal{B},\mathcal{K})$, then we shall say that a sequence $\{f_n\}_{n\in\mathbb{N}}$ of real-valued functions defined on $[0, 1]$ converges in category to a function $f\colon [0, 1] \to \mathbb{R}$ if it converges to f with respect to $\mathcal{I}$.

Theorem 2.16 *Convergence in category does not generate a Frechet topology in the family of functions having the Baire property.*

Proof This is a consequence of Theorems 1.20 and 2.14. □

Remark 2.17 The following simple example shows the same as Theorem 2.16. For the convenience all functions will be defined on $[0, 2]$. Let $f_n = \chi_{[1,1+\frac{1}{n}]}$ for $n \in \mathbb{N}$ and $f_{n,k} = \chi_{E_{n,k}}$ for $n \in \mathbb{N}$, where

$$E_{n,k} = \left[1, 1 + \frac{1}{n}\right] \cup \bigcup_{j=1}^{n-1}\left(\frac{j}{n} - \frac{1}{2^k}, \frac{j}{n} + \frac{1}{2^k}\right).$$

Put $A = \{f_{n,k} \colon n, k \in \mathbb{N}\}$. Then $\{f_n \colon n \in \mathbb{N}\} \subset \overline{A}$, where $\overline{A}$ is the "closure" of A. Obviously $\overline{\overline{A}}$ includes the function $f \equiv 0$.

Suppose that there exists a sequence $\{f_{n_m,k_m}\}_{m\in\mathbb{N}}$ from A which converges to 0 $\mathcal{K}$-a.e. Then the set $\{n_m \colon m \in \mathbb{N}\}$ must be infinite, so it is possible to chose an increasing subsequence $\{n_{m_p}\}_{p\in\mathbb{N}}$. From the assumption it follows that $\{f_{n_{m_p},k_{m_p}}\}_{p\in\mathbb{N}}$ also converges to 0 $\mathcal{K}$-a.e. But $\limsup_{p\to\infty} E_{n_{m_p},k_{m_p}}$ is residual in $[0, 1]$, so $\{f_{n_{m_p},k_{m_p}}\}_{p\in\mathbb{N}}$ cannot converge to 0 $\mathcal{K}$-a.e.—a contradiction.

Remark 2.18 Observe that Theorem 2.14 concerns the case when the closure of the set of functions is defined as on p. 25 (before Theorem 2.14). It does not exclude the existence of a topology connected with the convergence in category. We shall return to this topic in Chap. 15 (Remark 15.35 and Definition 15.36).

Chapter 3
Cauchy Condition

This chapter will be devoted to the Cauchy condition for the convergence in measure and in category. The proof that a sequence of measurable functions converges in measure if and only if it fulfills the Cauchy condition (or is fundamental in measure) is relatively simple. In the case of category the situation is different. There is no a set function like measure, so the definition of the Cauchy condition (Definition 3.6) needs a reformulation. At the end there is an example of the pair $(\mathcal{S}, \mathcal{I})$ of σ-algebra and σ-ideal for which the convergence with respect to a σ-ideal is not equivalent to the Cauchy condition.

Definition 3.1 A sequence $\{f_n\}_{n\in\mathbb{N}}$ of real-valued measurable functions defined on $[0, 1]$ fulfills the Cauchy condition in measure (is fundamental in measure) if and only if

$$\lim_{n,m\to\infty} \lambda\left(\{x \in [0,1]\colon |f_n(x) - f_m(x)| > \varepsilon\}\right) = 0$$

for each $\varepsilon > 0$, which means that for each $\varepsilon > 0$ and $\eta > 0$ there exists $n_0 \in \mathbb{N}$ such that $\lambda\left(\{x \in [0,1]\colon |f_n(x) - f_m(x)| > \varepsilon\}\right) < \eta$ for each $n, m \geq n_0$.

Theorem 3.2 *If $\{f_n\}_{n\in\mathbb{N}}$ converges in measure to some function f, then it is fundamental in measure.*

Proof Take $\varepsilon > 0$ and $\eta > 0$. There exists $n_0 \in \mathbb{N}$ such that

$$\lambda\left(\left\{x \in [0,1]\colon |f_n(x) - f(x)| > \frac{\varepsilon}{2}\right\}\right) < \frac{\eta}{2}$$

and also

$$\lambda\left(\left\{x \in [0,1]\colon |f_m(x) - f(x)| > \frac{\varepsilon}{2}\right\}\right) < \frac{\eta}{2}$$

for $n, m \geq n_0$. Then $\lambda\left(\{x \in [0,1]\colon |f_n(x) - f_m(x)| > \varepsilon\}\right) < \eta$ for each $n, m \geq n_0$. □

W. Wilczyński, *Convergence in Measure and in Category*, Lecture Notes in Mathematics 2394, https://doi.org/10.1007/978-3-032-23143-7_3

Theorem 3.3 (Riesz) *If $\{f_n\}_{n\in\mathbb{N}}$ is fundamental in measure, then it converges in measure to some function f.*

Proof For each $k \in \mathbb{N}$ there exists $n_k \in \mathbb{N}$ such that

$$\lambda\left(\{x \in [0,1] \colon |f_n(x) - f_m(x)| > 2^{-k}\}\right) < 2^{-k}$$

for $n, m \geq n_k$. We can assume that a sequence $\{n_k\}_{k\in\mathbb{N}}$ is increasing. Put $A_k = \{x \in [0,1] \colon |f_{n_{k+1}}(x) - f_{n_k}(x)| > 2^{-k}\}$ and $B = \limsup_k A_k = \bigcap_{n=1}^{\infty} \bigcup_{k=n}^{\infty} A_k$. We have $\lambda(B) = 0$ and we shall show that the subsequence $\{f_{n_k}\}_{k\in\mathbb{N}}$ converges pointwise on $[0,1] \setminus B$. Indeed, if $x \in [0,1] \setminus B = \bigcup_{n=1}^{\infty}\left([0,1] \setminus \bigcup_{k=n}^{\infty} A_k\right)$, then there exists $n_0 \in \mathbb{N}$ such that $x \in [0,1] \setminus A_k$ for each $k \geq n_0$. Hence $|f_{n_{k+1}}(x) - f_{n_k}(x)| \leq 2^{-k}$ for $k \geq n_0$ and the series $f_{n_1}(x) + \sum_{k=1}^{\infty}\left(f_{n_{k+1}}(x) - f_{n_k}(x)\right)$ is convergent (even absolutely) to some finite number $f(x)$. But the k-th partial sum of this series equals f_{n_k}, so $\lim_{k\to\infty} f_{n_k}(x) = f(x)$ for each $x \in [0,1] \setminus B$ (which means almost everywhere). Moreover $|f(x) - f_{n_k}(x)| \leq \sum_{i=k}^{\infty}\left|f_{n_{i+1}}(x) - f_{n_i}(x)\right| \leq 2^{-k+1}$ for $x \in [0,1] \setminus \bigcup_{i=k}^{\infty} A_i$, so $\{x \in [0,1] \colon |f_{n_k}(x) - f(x)| > 2^{-k+1}\} \subset \bigcup_{i=k}^{\infty} A_i$ for each $k \in \mathbb{N}$ and we have

$$\lambda\left(\{x \in [0,1] \colon |f_{n_k}(x) - f(x)| > 2^{-k+1}\}\right) < 2^{-k+1}$$

which means that $\{f_{n_k}\}_{k\in\mathbb{N}}$ converges in measure to f.

Take $\varepsilon > 0$ and $\eta > 0$. There exists $\hat{n} \in \mathbb{N}$ such that

$$\lambda\left(\left\{x \in [0,1] \colon |f_{n_k}(x) - f(x)| > \frac{\varepsilon}{2}\right\}\right) < \frac{\eta}{2}$$

for $n_k \geq \hat{n}$ and

$$\lambda\left(\left\{x \in [0,1] \colon |f_n(x) - f_{n_k}(x)| > \frac{\varepsilon}{2}\right\}\right) < \frac{\eta}{2}$$

for $n, n_k \geq \hat{n}$. Since

$$\begin{aligned}\{x \in [0,1] \colon |f_n(x) - f(x)| > \varepsilon\} \subset &\left\{x \in [0,1] \colon |f_n(x) - f_{n_k}(x)| > \frac{\varepsilon}{2}\right\} \\ &\cup \left\{x \in [0,1] \colon |f_{n_k}(x) - f(x)| > \frac{\varepsilon}{2}\right\},\end{aligned}$$

finally we have $\lambda\left(\{x \in [0,1] \colon |f_n(x) - f(x)| > \varepsilon\}\right) < \eta$ for $n \geq \hat{n}$. □

Recall that a sequence $\{f_n\}_{n\in\mathbb{N}}$ of real measurable functions defined on $[0,1]$ converges in measure to f if and only if $\lim_{n\to\infty} \| f_n - f \| = 0$ (Theorem 2.10). It is natural to define Cauchy condition in terms of the functional $\| \cdot \|$.

Theorem 3.4 *A sequence $\{f_n\}_{n\in\mathbb{N}}$ is fundamental in measure if and only if for each $\varepsilon > 0$ there exists $n_0 \in \mathbb{N}$ such that $\| f_n - f_m \| < \varepsilon$ for each $n, m \geq n_0$.*

Proof Necessity. Take $\varepsilon = \eta > 0$. There exists $n_0 \in \mathbb{N}$ such that

$$\lambda\left(\{x \in [0,1]\colon |f_n(x) - f_m(x)| > \varepsilon\}\right) < \varepsilon,$$

so $\| f_n - f_m \| \leq \varepsilon$ for $n, m \geq n_0$.
Sufficiency. For each $\varepsilon > 0$ there exists $n_0 \in \mathbb{N}$ such that $\| f_n - f_m \| < \varepsilon$ for $n, m \geq n_0$. Hence

$$\lambda\left(\{x \in [0,1]\colon |f_n(x) - f_m(x)| > \varepsilon\}\right) < \varepsilon.$$

If $\varepsilon \leq \eta$, then obviously

$$\lambda\left(\{x \in [0,1]\colon |f_n(x) - f_m(x)| > \varepsilon\}\right) < \eta.$$

If $\varepsilon > \eta$, then there exists $n_0 \in \mathbb{N}$ such that $\| f_n - f_m \| < \eta$ for $n, m \geq n_0$. Hence

$$\lambda\left(\{x \in [0,1]\colon |f_n(x) - f_m(x)| > \eta\}\right) < \eta$$

and the result follows from the inclusion

$$\{x \in [0,1]\colon |f_n(x) - f_m(x)| > \varepsilon\} \subset \{x \in [0,1]\colon |f_n(x) - f_m(x)| > \eta\}.$$

□

Theorem 3.5 *A sequence $\{f_n\}_{n\in\mathbb{N}}$ is fundamental in measure if and only if for all increasing sequences $\{n_k\}_{k\in\mathbb{N}}$ and $\{m_k\}_{k\in\mathbb{N}}$ of positive integers the sequence $\{f_{n_k} - f_{m_k}\}_{k\in\mathbb{N}}$ converges in measure to a function which is a.e. equal to zero.*

Proof Suppose first that $\{f_n\}_{n\in\mathbb{N}}$ is fundamental in measure, which is equivalent that it converges in measure. It is nearly obvious that $\{f_{n_k} - f_{m_k}\}_{k\in\mathbb{N}}$ converges in measure to zero for all increasing sequences $\{n_k\}_{k\in\mathbb{N}}$ and $\{m_k\}_{k\in\mathbb{N}}$.

Suppose that $\{f_n\}_{n\in\mathbb{N}}$ is not fundamental in measure. It means that there exist $\varepsilon > 0$ and $\eta > 0$ such that for each $k \in \mathbb{N}$ there exist $n_k, m_k \geq k$ for which $\lambda\left(\{x \in [0,1]\colon |f_{n_k}(x) - f_{m_k}(x)| > \varepsilon\}\right) \geq \eta$, so $\{f_{n_k} - f_{m_k}\}_{k\in\mathbb{N}}$ does not converges in measure to zero. □

The last theorem gives the possibility to introduce a kind of Cauchy condition for the convergence in category.

Definition 3.6 We shall say that a sequence $\{f_n\}_{n\in\mathbb{N}}$ fulfills the Cauchy condition in category if and only if for all increasing sequences $\{n_k\}_{k\in\mathbb{N}}$ and $\{m_k\}_{k\in\mathbb{N}}$ of positive integers the sequence $\{f_{n_k} - f_{m_k}\}_{k\in\mathbb{N}}$ converges in category to a function which is $\mathcal{K}$-a.e. equal to zero.

Suppose that $\{f_n\}_{n\in\mathbb{N}}$ is a sequence of functions having the Baire property defined on $[0, 1]$ which does not converge in category. Then there exists a subsequence $\{f_{n_m}\}_{m\in\mathbb{N}}$ of $\{f_n\}_{n\in\mathbb{N}}$ such that each subsequence $\{f_{n_{m_p}}\}_{p\in\mathbb{N}}$ does nor converge $\mathcal{K}$-almost everywhere. We shall look more carefully what happens in this situation.

To avoid complicated indices we shall use self-explaining denotations: $\{f_n\}_{n\in M}$ and $\{f_n^\alpha\}_{n\in\mathbb{N}}$, where α is an ordinal number.

Lemma 3.7 *If $\{f_n\}_{n\in\mathbb{N}}$ does not converge in category, then there exists an infinite set $M_1 \subset \mathbb{N}$ such that the sequence $\{f_n\}_{n\in M_1}$ has the following properties:*

(1) $Y = \{x \in [0, 1]\colon \limsup_{n\in M_1} f_n(x) - \liminf_{n\in M_1} f_n(x) > 0\} \notin \mathcal{K}$,
(2) *for each infinite set $P \Subset M_1$ if $B = \{x \in [0, 1]\colon \limsup_{n\in P} f_n(x) - \liminf_{n\in P} f_n(x) > 0\}$, then $Y \setminus B \in \mathcal{K}$ (observe that $B \subset Y$).*

Proof Let $\{f_n^{(1)}\}_{n\in\mathbb{N}}$ be a subsequence of $\{f_n\}_{n\in\mathbb{N}}$ without subsequences convergent $\mathcal{K}$-almost everywhere. If $\{f_n^{(1)}\}_{n\in\mathbb{N}}$ fulfills (1) and (2), the proof is finished. If not, then there exists a subsequence $\{f_n^{(2)}\}_{n\in\mathbb{N}}$ of $\{f_n^{(1)}\}_{n\in\mathbb{N}}$ such that

$$\{x \in [0, 1]\colon \limsup_{n\to\infty} f_n^{(1)}(x) - \liminf_{n\to\infty} f_n^{(1)}(x) > 0\}\setminus$$
$$\setminus\{x \in [0, 1]\colon \limsup_{n\in M_1} f_n^{(2)}(x) - \liminf_{n\in M_1} f_n^{(2)}(x) > 0\} \notin \mathcal{K}.$$

Suppose that we have defined subsequences $\{f_n^{(\beta)}\}_{n\in\mathbb{N}}$ for $\beta < \alpha$, $\alpha < \omega_1$ such that if $\gamma < \beta < \alpha$, then $\{f_n^{(\beta)}\}_{n\in\mathbb{N}}$ is almost a subsequence of $\{f_n^{(\gamma)}\}_{n\in\mathbb{N}}$ and

$$\{x \in [0, 1]\colon \limsup_{n\to\infty} f_n^{(\gamma)}(x) - \liminf_{n\to\infty} f_n^{(\gamma)}(x) > 0\}\setminus$$
$$\setminus\{x \in [0, 1]\colon \limsup_{n\to\infty} f_n^{(\beta)}(x) - \liminf_{n\to\infty} f_n^{(\beta)}(x) > 0\} \notin \mathcal{K}.$$

If α has a predecessor, $\alpha = \beta + 1$ and $\{f_n^{(\beta)}\}_{n\in\mathbb{N}}$ does not fulfill 1) and 2), we choose a subsequence $\{f_n^{(\alpha)}\}_{n\in\mathbb{N}}$ of $\{f_n^{(\beta)}\}_{n\in\mathbb{N}}$ such that

$$\{x \in [0, 1]\colon \limsup_{n\to\infty} f_n^{(\beta)}(x) - \liminf_{n\to\infty} f_n^{(\beta)}(x) > 0\}\setminus$$
$$\setminus\{x \in [0, 1]\colon \limsup_{n\to\infty} f_n^{(\alpha)}(x) - \liminf_{n\to\infty} f_n^{(\alpha)}(x) > 0\} \notin \mathcal{K}.$$

If α is a limit ordinal, then we choose an increasing sequence $\{\beta_n\}_{n\in\mathbb{N}}$ such that $\lim_{n\to\infty} \beta_n = \alpha$ and put $f_n^{(\alpha)} = f_n^{(\beta_n)}$. The procedure must stop since $(\mathcal{B}, \mathcal{K})$ fulfills the countable chain condition. □

Lemma 3.8 *Under the assumptions and denotations from the previous lemma there exists an infinite subset $M_2 \subset M_1$ such that for each infinite set $P \subset M_2$*

$$\{x \in [0, 1]\colon \limsup_{n\in P} f_n(x) - \liminf_{n\in P} f_n(x) < \limsup_{n\in M_2} f_n(x) - \liminf_{n\in M_2} f_n(x)\} \in \mathcal{K}.$$

Proof The proof goes similarly using the transfinite induction. □

Lemma 3.9 *Under the assumptions and denotations from the previous lemmas there exist: a set $Y \in [0, 1]$ of the second category and having the Baire property, a pair of numbers $r \in \mathbb{R}$ and $\delta > 0$, and an infinite set $M \subset M_2$ such that for each infinite set $L \Subset M$ and for each set $A \subset Y$, $A \in \mathcal{B} \setminus \mathcal{K}$, there exist a pair of points $x, y \in A$ and an infinite set $L_0 \subset L$ such that $f_n(x) > r + \delta$ and $f_n(y) < r$ for each $n \in L_0$.*

Proof Let

$$Y_{r,\delta} = \{x \in [0, 1]\colon \limsup_{n \in M_2} f_n(x) > r + \delta \text{ and } \liminf_{n \in M_2} f_n(x) < r\}$$

for $r \in \mathbb{Q}$ and $\delta \in \mathbb{Q}$, $\delta > 0$. Then

$$\bigcup_{r,\delta} Y_{r,\delta} = \{x \in [0, 1]\colon \limsup_{n \in M_2} f_n(x) - \liminf_{n \in M_2} f_n(x) > 0\},$$

so there exists a pair $r_0 \in \mathbb{Q}$ and $\delta_0 \in \mathbb{Q}$, $\delta_0 > 0$ such that $Y_{r_0,\delta_0} \notin \mathcal{K}$. Put $Y = Y_{r_0,\delta_0}$. □

Let $\mathcal{M}$ denote the family of all real functions, $\mathcal{S}$-measurable, $\mathcal{I}$-a.e. finite and $\mathcal{I}$-a.e. defined on $[0, 1]$.

Lemma 3.10 *([W-BW99]) Let $\{f_n\}_{n \in \mathbb{N}}$ be a sequence of functions from $\mathcal{M}$ such that each subsequence $\{f_{n_m}\}_{m \in \mathbb{N}}$ of $\{f_n\}_{n \in \mathbb{N}}$ diverges on the set of the second category. Then there exist two increasing sequences $\{n_k\}_{k \in \mathbb{N}}$ and $\{m_k\}_{k \in \mathbb{N}}$ of natural numbers such that $\{f_{n_k} - f_{m_k}\}_{k \in \mathbb{N}}$ does not converge to zero in category.*

Proof Let $P_n \subset [0, 1]$ for $n \in \mathbb{N}$ be a set of the first category such that $f_n|[0,1] - P_n$ is continuous (compare [Oxt80], Theorem 8.1).

Let Y, r, δ and M be as in Lemma 3.9. We can obviously suppose that $Y \subset [0, 1] - \bigcup_{n=1}^{\infty} P_n$, so all functions are continuous at each point of Y. Let $(a, b) \subset [0, 1]$ be an interval such that $(a, b) - Y \in \mathcal{I}$ (recall that $Y \in \mathcal{S} - \mathcal{I}$). Observe that if $(c, d) \subset (a, b)$, then also $(c, d) - Y \in \mathcal{I}$.

Put $A = (a, b) \cap Y$. From Lemma 3.9 there exist an infinite set $M_1^1 \subset M$ and a point $x_1^1 \in A$ such that $f_n(x_1^1) > r + \delta$ for $n \in M_1^1$. Choose $n_1 \in M_1^1$. From the continuity of f_{n_1} at x_1^1 on Y it follows that there exists $\eta_1 > 0$ such that $f_{n_1}(x) > r + \delta$ for $x \in (x_1^1 - \eta_1, x_1^1 + \eta_1) \cap Y$. Now for $A = (x_1^1 - \eta_1, x_1^1 + \eta_1) \cap Y$ and M_1^1 again by virtue of Lemma 3.9 there exists an infinite set $M_2^1 \subset M_1^1$ and a point $y_1^1 \in A$ such that $f_n(y_1^1) < r$ for $n \in M_2^1$. Choose $m_1 \in M_2^1$. From the continuity of f_{m_1} at y_1^1 on Y if follows that there exists $\varepsilon_1 > 0$ such that $f_{m_1}(x) < r$ for $x \in (y_1^1 - \varepsilon_1, y_1^1 + \varepsilon_1) \cap Y \subset (x_1^1 - \eta_1, x_1^1 + \eta_1) \cap Y$. Let $G_1 = (y_1^1 - \varepsilon_1, y_1^1 + \varepsilon_1)$. Then we have $f_{n_1}(x) - f_{m_1}(x) > \delta$ on $G_1 \cap Y$.

Suppose that we have chosen increasing finite sequences $(n_1, \dots, n_{k-1})$ and $(m_1, \dots, m_{k-1})$ and a sequence $(G_1, \dots, G_{k-1})$ of open sets such that

$$G_j \cap \left(a + \frac{i-1}{j}(b-a), a + \frac{i}{j}(b-a)\right) \neq \varnothing$$

for $j \in \{1, \dots, k-1\}$ and $i \in \{1, \dots, j\}$ and $f_{n_j}(x) - f_{m_j}(x) > \delta$ for $x \in G_j \cap Y$, $j \in \{1, \dots, k-1\}$. We have also chosen descending sequences of infinite sets of natural numbers $M_1^1 \supset M_2^1; M_1^2 \supset \dots \supset M_4^2; \dots; M_1^{k-1} \supset \dots \supset M_{2(k-1)}^{k-1}$.

To make an inductive step we divide (a, b) into k parts $\{(a_i^k, b_i^k)\}_{i=1,\dots,k}$ of equal length, i.e. such that $a_i^k = a + \frac{i-1}{k}(b-a), b_i^k = a + \frac{i}{k}(b-a)$. For $A = (a_1^k, b_1^k) \cap Y$ and M we use Lemma 3.9 to find an infinite set $M_1^k \subset M$ and a point $x_1^k \in A$ such that $f_n(x_1^k) > r + \delta$ for $n \in M_1^k$. Then we repeat the procedure for $A = (a_2^k, b_2^k) \cap Y$ and M_1^k obtaining $x_2^k \in A$ and an infinite set $M_2^k \subset M_1^k$ such that $f_n(x_2^k) > r + \delta$ for $n \in M_2^k$ (observe that also $f_n(x_1^k) > r + \delta$ for $n \in M_2^k$). After k choices we obtain an infinite set M_k^k of natural numbers and a set $\{x_1^k, \dots, x_k^k\} \subset Y$ such that $x_1^k \in (a_i^k, b_i^k)$ and $f_n(x_i^k) > r + \delta$ for $i \in \{1, \dots, k\}$ and $n \in M_k^k$.

Choose $n_k \in M_k^k$ such that $n_k > n_{k-1}$. From the continuity of f_{n_k} at each x_i^k, $i \in \{1, \dots, k\}$, on Y it follows that there exists $\eta_k > 0$ such that $f_{n_k}(x) > r + \delta$ for $x \in \bigcup_{i=1}^k (x_i^k - \eta_k, x_i^k + \eta_k) \cap Y$.

To find m_k and the open set G_k we proceed as follows: for $A = (x_1^k - \eta_1, x_1^k + \eta_1) \cap (a_1^k, b_1^k) \cap Y$ and M_k^k applying Lemma 3.9 we choose $y_1^k \in A$ and an infinite set $M_{k+1}^k \subset M_k^k$ such that $f_n(y_1^k) < r$ for $n \in M_{k+1}^k$. Then we choose successively infinite sets of natural numbers $M_{k+2}^k \supset M_{k+3}^k \supset \dots \supset M_{2k}^k$ (of course, $M_{k+2}^k \subset M_{k+1}^k$) and points $y_2^k, \dots, y_k^k$ such that $y_i^k \in (x_i^k - \eta_k, x_i^k + \eta) \cap (a_i^k, b_i^k) \cap Y$ for $i \in \{2, \dots, k\}$ and $f_n(y_i^k) < r$ for $n \in M_{k+i}^k$. At least, we choose $m_k > m_{k-1}$ from the set M_{2k}^k. Then we have $f_{m_k}(y_i^k) < r$ for $i \in \{1, \dots, k\}$. From the continuity of f_{m_k} at each y_i^k, $i \in \{1, \dots, k\}$, it follows that there exists $\varepsilon_k > 0$ such that $(y_i^k - \varepsilon_k, y_i^k + \varepsilon_k) \subset (x_i^k - \eta_k, x_i^k + \eta)$ and if we put $G_k = \bigcup_{i=1}^k (y_i^k - \varepsilon_k, y_i^k + \varepsilon_k)$, then $f_{m_k}(x) < r$ for $x \in G_k \cap Y$. Then we have $f_{n_k}(x) - f_{m_k}(x) > \delta$ for $x \in G_k \cap Y$ and $G_k \cap \left(a + \frac{i-1}{k}(b-a), \frac{i}{k}(b-a)\right) \neq \varnothing$ for $i = \{1, \dots, k\}$.

Since for each increasing sequence $\{k_p\}_{p \in \mathbb{N}}$ of natural numbers we have $E = \{x \in [0,1] : \limsup_p (f_{n_{k_p}}(x) - f_{m_{k_p}}(x)) \geq \delta\} \supset \bigcap_{j=1}^\infty \bigcup_{p=j}^\infty \{x \in [0,1] : f_{n_{k_p}}(x) - f_{m_{k_p}}(x) > \delta\} \supset Y \cap \bigcap_{j=1}^\infty \bigcup_{p=j}^\infty G_{k_p}$ and, as is not difficult to see, the set $\bigcup_{p=j}^\infty G_{k_p}$ is open and dense in (a, b) for each $j \in \mathbb{N}$, so E is residual in (a, b) and hence $\{f_{n_{k_p}} - f_{m_{k_p}}\}_{p \in \mathbb{N}}$ does not converge $\mathcal{I}$-a.e.. Finally, $\{f_{n_k} - f_{m_k}\}_{k \in \mathbb{N}}$ does not converge in category. □

Theorem 3.11 *([W-BW99]) If a sequence $\{f_n\}_{n \in \mathbb{N}}$ of functions from $\mathcal{M}$ is fundamental in category, then it converges in category to some function $f \in \mathcal{M}$.*

Proof Suppose that $\{f_n\}_{n \in \mathbb{N}}$ does not converge in category to any function from $\mathcal{M}$. Then, either (1) there exists a pair of functions $g_1, g_2 \in \mathcal{M}$ which are not equivalent and a pair of increasing sequences $\{n_k\}_{k \in \mathbb{N}}$ and $\{m_k\}_{k \in \mathbb{N}}$ of natural numbers such that $f_{n_k} \xrightarrow[k\to\infty]{} g_1$ $\mathcal{I}$-a.e. and $f_{m_k} \xrightarrow[k\to\infty]{} g_2$ $\mathcal{I}$-a.e., or (2) there exists an increasing sequence $\{n_m\}_{m \in \mathbb{N}}$ of natural numbers such that for each subsequence $\{n_{m_p}\}_{p \in \mathbb{N}}$ of $\{n_m\}_{m \in \mathbb{N}}$ the sequence $\{f_{n_{m_p}}\}_{p \in \mathbb{N}}$ diverges on the set of the second category, or (3) there exists an increasing sequence $\{n_m\}_{m \in \mathbb{N}}$ of natural numbers such that $\{x \in [0,1] : \lim_{m\to\infty} f_{n_m}(x) = +\infty\} \notin \mathcal{I}$ or $\{x \in [0,1] : \lim_{m\to\infty} f_{n_m}(x) = -\infty\} \notin \mathcal{I}$. In case (1) the sequence $\{f_n\}_{n \in \mathbb{N}}$ is not fundamental in category because $\{f_{n_k} - f_{m_k}\}_{k \in \mathbb{N}}$ does not converge to zero in category. In the case (2) from Lemma 3.10 we obtain immediately that $\{f_n\}_{n\in\mathbb{N}}$ is not fundamental in category. Suppose now that in case (3) the first possibility holds. To simplify the description assume that $\{n_m\}_{m\in\mathbb{N}}$ is the sequence $\{n\}_{n\in\mathbb{N}}$ of all natural numbers, i.e. that $E = \{x \in [0,1] : \lim_{n\to\infty} f_n(x) = +\infty\} \notin \mathcal{I}$. Since $E = \bigcap_{k=1}^{\infty}\bigcup_{m=1}^{\infty}\bigcap_{n=m}^{\infty}\{x \in [0,1] : f_n(x) \geq k\}$, the set E has the Baire property and there exists an interval (a,b) such that $(a,b) - E \in \mathcal{I}$. Take $Y = (a,b) \cap E - \bigcup_{n=1}^{\infty} P_n$, where, as before, $P_n \in \mathcal{I}$ is a set such that $f_n|[0,1] - P_n$ is continuous. We shall find an increasing sequence $\{j_n\}_{n\in\mathbb{N}}$ of natural numbers such that $\{f_{j_n} - f_n\}_{n\in\mathbb{N}}$ does not converge to zero in category. Put $j_1 = 1$. Suppose that we have constructed j_{n-1}. To find j_n choose an increasing sequence $x_1^n < x_2^n < \cdots < x_{k_n}^n$, such that $x_i^n \in Y$ for $i \in \{1,\ldots,k_n\}$, $x_1^n - a < \frac{1}{n}$, $b - x_{k_n}^n < \frac{1}{n}$ and $x_i^n - x_{i-1}^n < \frac{1}{n}$ for $i \in \{2,\ldots,k_n\}$.

Let $c_n = \max\{f_n(x_1^n),\ldots,f_n(x_{k_n}^n)\}$. Since $\{f_n(x)\}_{n\in\mathbb{N}}$ diverges to $+\infty$ at each point $x \in Y$, there exists a natural number $j_n > j_{n-1}$ such that

$$\min\{f_{j_n}(x_1^n),\ldots,f_{j_n}(x_{k_n}^n)\} > c_n + 1.$$

From the continuity of all f_n's on Y we conclude that there exists on open set $G_n \supset \{x_1^n,\ldots,x_{k_n}^n\}$ such that $f_{j_n}(x) - f_n(x) > 1$ for all $x \in G_n \cap Y$. Hence for each increasing sequence $\{n_p\}_{p\in\mathbb{N}}$ of natural numbers we have $\{x \in (a,b) : \{f_{j_{n_p}}(x) - f_{n_p}(x)\}_{p\in\mathbb{N}}$ does not converge to zero$\} \supset Y \cap \limsup_p G_{n_p}$. It is not difficult to see that the last set is residual in (a,b). Hence $\{f_{j_n} - f_n\}_{n\in\mathbb{N}}$ does not converge to zero in category, so $\{f_n\}_{n\in\mathbb{N}}$ is not fundamental in category. □

Let $(\mathcal{S},\mathcal{I})$ be a measurable space and $\mathcal{I} \subset \mathcal{S}$—a proper σ-ideal. We shall say that a sequence $\{f_n\}_{n\in\mathbb{N}}$ of $\mathcal{S}$-measurable real-valued functions defined on X is fundamental with respect to $\mathcal{I}$ if and only if for each increasing sequences of positive integers $\{n_i\}_{i\in\mathbb{N}}$ and $\{m_i\}_{i\in\mathbb{N}}$ there exists an increasing sequence $\{i_j\}_{j\in\mathbb{N}}$ of positive integers such that the sequence $\{f_{n_{i_j}} - f_{m_{i_j}}\}_{j\in\mathbb{N}}$ converges $\mathcal{I}$-a.e. to zero.

In the sequel (based on [Wla60]) T will denote the set of all increasing sequences of positive integers. If $\mathfrak{n} = \{n_i\}_{i\in\mathbb{N}}$ and $\mathfrak{m} = \{m_i\}_{i\in\mathbb{N}}$ are elements of T we shall write $\mathfrak{m} \prec \mathfrak{n}$ if and only if there exists $k_0 \in \mathbb{N}$ such that $m_i \leq n_i$ for $i \geq k_0$. The relation $\prec$ is a partial order in T. The set $T_0 \subset T$ is bounded from above if there exists $\mathfrak{n} \in T$ such that $\mathfrak{m} \prec \mathfrak{n}$ for each $\mathfrak{m} \in T_0$. Let $\mathfrak{b} = \min\{\mathrm{card}(T_0) \colon T_0 \subset T \text{ is unbounded}\}$. It is well known that $\aleph_0 < \mathfrak{b} \leq 2^{\aleph_0}$. It is also well known that $\mathfrak{b}$ is a regular cardinal number, which means that a set of cardinality $\mathfrak{b}$ is not the union of less than $\mathfrak{b}$ sets of cardinality $< \mathfrak{b}$.

Lemma 3.12 *Suppose that* $\mathfrak{n} = \{n_i\}_{i\in\mathbb{N}}$ *and* $\mathfrak{m} = \{m_i\}_{i\in\mathbb{N}}$ *are increasing sequences of positive integers such that* $\mathfrak{m} \prec \mathfrak{n}$. *Put* $n(1) = n_1 + 1$ *and* $n(i+1) = n(i) + n_{n(i)}$, $m(1) = m_1 + 1$ *and* $m(i+1) = m(i) + m_{m(i)}$. *Let* $j = j(i)$ *be the smallest positive integer such that* $m(i) \leq n(j(i))$. *Then* $m(i+1) \leq n(j(i)+1)$ *for sufficiently big* i.

Proof There exists $i_0 \in \mathbb{N}$ such that $m_i < n_i$ for $i \geq i_0$. Observe that $m(i) > i$ for each $i \in \mathbb{N}$. Hence for $i \geq i_0$ we have

$$m(i+1) = m(i) + m_{m(i)} \leq m(i) + n_{m(i)} \leq n(j(i)) + n_{m(i)} \leq$$
$$\leq n(j(i)) + n_{n(j(i))} = n(j(i)+1).$$

□

Lemma 3.13 *Suppose that $T_0 \subset T$ and $\operatorname{card}(T_0) < \mathbb{b}$. Then there exists a sequence $\{c_j\}_{j\in\mathbb{N}}$ of non-negative real numbers such that*

a) $\sum_{j=1}^{\infty} c_j = \infty$,

b) for each $\mathbb{m} \in T_0$ if $\sigma_k(\mathbb{m}) = \sum_{j=k+1}^{k+m_k} c_j$, then $\lim_{k\to\infty} \sigma_k(\mathbb{m}) = 0$,

Proof T_0 is bounded from above, so there exists $\mathbb{n} = \{n_i\}_{i\in\mathbb{N}}$ in T such that $\mathbb{m} \prec \mathbb{n}$ for each $\mathbb{m} \in T_0$. Put $m(1) = m_1 + 1$, $m(i+1) = m(i) + m_{m(i)}$ and $n(1) = n_1 + 1$, $n(i+1) = n(i) + n_{n(i)}$. Let $c_j = \frac{1}{i}$ if $j = n(i)$ and $c_j = 0$ for remaining j's. Then from Lemma 3.12 it follows that for sufficiently big i among numbers $c_{m(i)+1}, c_{m(i)+2}, \ldots, c_{m(i)+m_{m(i)}}$ there is only one different from zero, which proves b). The condition a) is obviously fulfilled. □

Let A be the set of all ordinal numbers less than $\omega_{\mathbb{b}}$, where $\omega_{\mathbb{b}}$ is the smallest ordinal of cardinality $\mathbb{b}$ and let X_0 be the set of all real-valued functions defined on A. Put $\mathcal{S} = 2^A$. Then all functions from X_0 are measurable. At last let $\mathcal{I} = \{B \subset A : \operatorname{card}(B) < \mathbb{b}\}$. The family $\mathcal{I} \subset \mathcal{S}$ is a proper σ-ideal, since $\mathbb{b}$ is regular.

Theorem 3.14 ([Wla60]) *There exists a sequence $\{f_n\}_{n\in\mathbb{N}}$ of functions from X_0 which is fundamental with respect to $\mathcal{I}$ and does not converge to any function which is finite $\mathcal{I}$-a.e.*

Proof We shall use the denotation $\mathbb{n}^\alpha = \{n_i^\alpha\}_{i\in\mathbb{N}}$. Let $T_0 = \{\mathbb{n}^\alpha \in T : \alpha < \omega_{\mathbb{b}}\}$ be an unbounded set (w.r.t. $\prec$). For each $\alpha \in A$ using Lemma 3.13 we shall construct a sequence $\{c_m(\alpha)\}_{m\in\mathbb{N}}$ of non-negative real numbers such that

(a) $\sum_{m=1}^{\infty} c_m(\alpha) = \infty$,

(b) if $\beta < \alpha$, then $\lim_{k\to\infty} = \sum_{m=k+1}^{k+n_k^\beta} c_m(\alpha) = 0$.

For $m \in \mathbb{N}$ and $\alpha \in A$ put $f_m(\alpha) = c_m(\alpha)$. Let $F_m = \sum_{k=1}^{m} f_k$. It is obvious that $\lim_{m\to\infty} F_m(\alpha) = \infty$ for each $\alpha \in A$. We shall prove that the sequence $\{F_m\}_{m\in\mathbb{N}}$ is fundamental with respect to $\mathcal{I}$. Take arbitrary increasing sequences $\{n_i\}_{i\in\mathbb{N}}$ and $\{m_i\}_{i\in\mathbb{N}}$ of positive integers. We can assume that $m_i < n_i$ for $i \in \mathbb{N}$. Put $p_i = n_i - m_i$. We shall prove that the sequence $\left\{\sum_{n=m_k+1}^{m_k+p_k} f_n\right\}_{n\in\mathbb{N}}$ includes a subsequence convergent to zero $\mathcal{I}$-almost everywhere. Put $r_k = \max_{i\leq k}(m_i + p_i)$. Taking into account that $\{m_i\}_{i\in\mathbb{N}}$ is increasing we have $m_{k+1} \geq k$. Simultaneously $m_k + p_k \leq k + r_k$. Hence

$$\sum_{i=m_k+1}^{m_k+p_k} f_i(\alpha) = \sum_{i=m_k+1}^{m_k+p_k} c_i(\alpha) \leq \sum_{i=k+1}^{k+r_k} c_i(\alpha).$$

There exists an ordinal number $\beta < \omega_{\mathbb{b}}$ such that it is not true that $\mathbb{n}^{\beta} \prec \mathbb{r}$, where $\mathbb{r} = \{r_k\}_{k\in\mathbb{N}}$, because T_0 is an unbounded set. It means that there exists an increasing sequence $\{k_j\}_{j\in\mathbb{N}}$ for which $r_{k_j} < n_{k_j}^{(\beta)}$ when $j \in \mathbb{N}$. Hence

$$0 \leq \sum_{i=m_k+1}^{k_j+r_{k_j}} c_i(\alpha) \leq \sum_{i=k_j+1}^{k_j+n_{k_j}^{(\beta)}} c_i(\alpha).$$

The sequence $\left\{\sum_{i=k_j+1}^{k_j+n_{k_j}^{(\beta)}} c_i(\alpha)\right\}_{j\in\mathbb{N}}$ converges to 0 when $j \to \infty$ for all $\alpha \geq \beta$. It follows from the construction of sequences $\{c_m(\alpha)\}_{m\in\mathbb{N}}$ for $\alpha < \omega_{\mathbb{b}}$. The sequence $\left\{\sum_{i=n_{k_j}+1}^{n_{k_j}+p_{k_j}} c_i(\alpha)\right\}_{j\in\mathbb{N}}$ also converges to 0 when $j \to \infty$ for all $\alpha \geq \beta$. Since $\mathrm{card}\{\gamma\colon \gamma < \beta\} < \mathbb{b}$, it means the convergence $\mathcal{I}$-almost everywhere. □

Chapter 4
Measurable Boundaries of an Arbitrary Function

The upper and lower measurable boundaries of arbitrary (not necessarily measurable) functions were studied by H. Blumberg in [Blu35] and C. Goffman and R. E. Zink in [GZ60]. Blumberg has defined boundaries at each point separately and the definition of Goffman and Zink is global. As a result both boundaries of Blumberg are unique while if f is an upper or lower boundary of some function in the sense of Goffman and Zink, and $f = g$ a.e., then g is also an upper or lower boundary of this function. In this chapter we shall study the measurable boundaries of Goffman and Zink, of Blumberg and connected with the density topology as well as their analogues for the Baire category. All considered functions will be defined on $[0, 1]$. Since the essential role in all definitions is played by the order relation in the range of functions, we shall suppose that all functions under consideration transform $[0, 1]$ into $[-1, 1]$.

It is well known that if $f\colon [0, 1] \to \mathbb{R}$ is measurable, then $L(f) = \{(x, y)\colon x \in [0, 1],\ y < f(x)\}$ and $U(f) = \{(x, y)\colon x \in [0, 1],\ y > f(x)\}$ belong to $\mathcal{L}_2$ and if $f\colon [0, 1] \to \mathbb{R}$ has the Baire property, then $L(f)$ and $U(f)$ belong to $\mathcal{B}_2$.

As a direct consequence of Fubini and Kuratowski-Ulam theorems we obtain the following lemmas:

Lemma 4.1 *If f_1, f_2 are measurable real-valued functions defined on $[0, 1]$ such that $f_1(x) = f_2(x)$ for $x \in [0, 1] \setminus A$ and $f_2(x) < f_1(x)$ for $x \in A$, where $A \in \mathcal{L} \setminus \mathcal{N}$, then $U(f_2) \setminus U(f_1) \in \mathcal{L}_2 \setminus \mathcal{N}_2$ and $L(f_2) \setminus L(f_1) \in \mathcal{L}_2 \setminus \mathcal{N}_2$.*

Lemma 4.2 *If f_1, f_2 are real-valued functions defined on $[0, 1]$ having the Baire property such that $f_1(x) = f_2(x)$ for $x \in [0, 1] \setminus A$ and $f_2(x) < f_1(x)$ for $x \in A$, where $A \in \mathcal{B} \setminus \mathcal{K}$, then $U(f_2) \setminus U(f_1) \in \mathcal{B}_2 \setminus \mathcal{K}_2$ and $L(f_2) \setminus L(f_1) \in \mathcal{B}_2 \setminus \mathcal{K}_2$.*

If f_1, f_2 are real-valued functions defined on $[0, 1]$ then $f_1 \leqslant f_2$ will mean that $\{x \in [0, 1]\colon f_1(x) > f_2(x)\}$ belongs to $\mathcal{N}$ or $\mathcal{K}$ according to the context.

Definition 4.3 If $\mathcal{A}$ is a family of measurable real-valued functions defined on $[0, 1]$, then we say that a measurable function $h\colon [0, 1] \to [-1, 1]$ is a lower bound in measure of $\mathcal{A}$ if $h \leqslant f$ for each $f \in \mathcal{A}$. We say that $g\colon [0, 1] \to [-1, 1]$ is a greatest

W. Wilczyński, *Convergence in Measure and in Category*, Lecture Notes in Mathematics 2394, https://doi.org/10.1007/978-3-032-23143-7_4

lower bound in measure of $\mathcal{A}$ ($g = \inf \mathcal{A}$) if g is a lower bound in measure of $\mathcal{A}$ and $h \leqslant g$ for each lower bound h in measure of $\mathcal{A}$. We say that a measurable function $h\colon [0,1] \to [-1,1]$ is an upper bound in measure of $\mathcal{A}$ if $f \leqslant h$ for each $f \in \mathcal{A}$. We say that $g\colon [0,1] \to [-1,1]$ is a least upper bound in measure of $\mathcal{A}$ ($g = \sup \mathcal{A}$) if g is an upper bound in measure of $\mathcal{A}$ and $g \leqslant h$ for each upper bound h in measure of $\mathcal{A}$.

Theorem 4.4 *For each non-empty family $\mathcal{A}$ of measurable functions defined on $[0,1]$ there exists* $\inf \mathcal{A}$ *and* $\sup \mathcal{A}$.

Proof Choose $g_0 \in \mathcal{A}$. If $g_0 = \inf \mathcal{A}$, the proof is finished. If it is not the case, then there exists $f_1 \in \mathcal{A}$ such that $A_0 = \{x \in [0,1]\colon g_0(x) > f_1(x)\} \in \mathcal{L} \setminus \mathcal{N}$. Put

$$g_1(x) = \begin{cases} g_0(x) & \text{for } x \in [0,1] \setminus A_0 \\ f_1(x) & \text{for } x \in A_0 \end{cases}.$$

Then $g_1 \leq g_0$ and $B_0 = \{(x,y)\colon x \in A_0,\ y \in (f_1(x), g_1(x)]\} = U(f_1) \setminus U(g_0) \in \mathcal{L}_2 \setminus \mathcal{N}_2$ (Lemma 4.1). Suppose that we have constructed g_β for $\beta < \alpha < \omega_1$. If α has a predecessor, $\alpha = \beta + 1$ and $g_\beta = \inf \mathcal{A}$, the proof is finished. If not, then there exists a function $f_\alpha \in \mathcal{A}$ such that $A_\beta = \{x \in [0,1]\colon g_\beta(x) > f_\alpha(x)\} \in \mathcal{L} \setminus \mathcal{N}$. Put

$$g_\alpha(x) = \begin{cases} g_\beta(x) & \text{for } x \in [0,1] \setminus A_\beta \\ f_\alpha(x) & \text{for } x \in A_\beta \end{cases}.$$

Then $g_\alpha \leq g_\beta$ and $B_\beta = \{(x,y)\colon x \in A_\beta,\ y \in (f_\alpha(x), g_\beta(x)]\} \in \mathcal{L}_2 \setminus \mathcal{N}_2$. If α is a limit number, then put $g_\alpha = \inf\{g_\beta\colon \beta < \alpha\}$. In both cases g_α is a measurable function. Observe also that if $\beta_1 < \beta_2 < \alpha$, then $B_{\beta_1} \cap B_{\beta_2} = \varnothing$. Since the pair $(\mathcal{L}_2, \mathcal{N}_2)$ fulfills the countable chain condition, there must be $\alpha_0 < \omega_1$ for which the procedure stops, so $g_{\alpha_0} = \inf \mathcal{A}$. The proof for $\sup \mathcal{A}$ is analogous. □

Remark 4.5 Observe that $\inf \mathcal{A}$ and $\sup \mathcal{A}$ are not unique. If $g = \inf \mathcal{A}$, then $g_1 = \inf \mathcal{A}$ if and only if $g = g_1$ a.e. and similarly for $\sup \mathcal{A}$.

Definition 4.6 ([GZ60]) We say that a real-valued measurable function G defined on $[0,1]$ is the upper measurable boundary of a function $f\colon [0,1] \to [-1,1]$ if and only if $f \leqslant G$ and for each measurable function $h\colon [0,1] \to [-1,1]$ if $f \leqslant h$ then $G \leqslant h$.

A real-valued measurable function g defined on $[0,1]$ is the lower measurable boundary of a function $f\colon [0,1] \to [-1,1]$ if and only if $g \leqslant f$ and for each measurable function $h\colon [0,1] \to [-1,1]$ if $h \leqslant f$ then $h \leqslant g$.

Theorem 4.7 *For arbitrary function $f\colon [0,1] \to [-1,1]$ there exists upper measurable boundary G of f and lower measurable boundary g of f.*

Proof Let $\mathcal{A}_1 = \{h\colon [0,1] \to [-1,1]\colon h \text{ is measurable and } f \leqslant h\}$ and $\mathcal{A}_2 = \{h\colon [0,1] \to [-1,1]\colon h \text{ is measurable and } h \leqslant f\}$. Then $G = \inf \mathcal{A}_1$ is an upper measurable boundary of f and $g = \sup \mathcal{A}_2$ is a lower measurable boundary of f.□

The notions of upper and lower measurable boundaries of an arbitrary function have a very satisfactory category analogues.

Definition 4.8 ([Wag81]) If $\mathcal{A}$ is a family of functions having the Baire property transforming $[0,1]$ into $[-1,1]$, then we say that a function $h\colon [0,1] \to [-1,1]$ having the Baire property is a lower bound in category of $\mathcal{A}$ if $h \leqslant f$ for each $f \in \mathcal{A}$.

We say that $g\colon [0,1] \to [-1,1]$ having the Baire property is a greatest lower bound in category of $\mathcal{A}$ ($g = \inf \mathcal{A}$) if g is a lower bound in category of $\mathcal{A}$ and $h \leqslant g$ for each lower bound h in category of $\mathcal{A}$.

We say that a function $h\colon [0,1] \to [-1,1]$ having the Baire property is an upper bound in category of $\mathcal{A}$ if $f \leqslant h$ for each $f \in \mathcal{A}$.

We say that $g\colon [0,1] \to [-1,1]$ having the Baire property is a least upper bound in category of $\mathcal{A}$ ($g = \sup \mathcal{A}$) if g is an upper bound in category of $\mathcal{A}$ and $g \leqslant h$ for each upper bound h in category of $\mathcal{A}$.

Theorem 4.9 *For each non-empty family $\mathcal{A}$ of functions having the Baire property transforming $[0,1]$ into $[-1,1]$ there exist $\inf \mathcal{A}$ and $\sup \mathcal{A}$.*

Proof The proof goes essentially the same way as the proof of Theorem 4.4. Lemma 4.2 and the fact that the pair $(\mathcal{B}_2, \mathcal{K}_2)$ fulfills the countable chain condition are used. □

Definition 4.10 ([Wag81]) We say that a function $G_c\colon [0,1] \to [-1,1]$ having the Baire property is the upper categorial boundary of a function $f\colon [0,1] \to [-1,1]$ if and only if $f \leqslant G_c$ and for each function $h\colon [0,1] \to [-1,1]$ having the Baire property if $f \leqslant h$ then $G_c \leqslant h$.

A function $g_c\colon [0,1] \to [-1,1]$ having the Baire property is the lower categorial boundary of a function $f\colon [0,1] \to [-1,1]$ if and only if $g_c \leqslant f$ and for each function $h\colon [0,1] \to [-1,1]$ having the Baire property if $h \leqslant f$ then $h \leqslant g_c$.

Theorem 4.11 *For arbitrary function $f\colon [0,1] \to [-1,1]$ there exists upper categorial boundary G_c of f and lower categorial boundary g_c of f.*

Proof Again the proof is essentially the same as that of Theorem 4.7 with the change of measurability onto the Baire property. □

Both functions G and g for measure and G_c and g_c for category were defined globally. Below we shall consider pointwise definitions of upper and lower measurable and categorial boundaries of an arbitrary function and we compare the unique obtained in this way function with the definitions of Blumberg, Goffman and Zink, and Wagner. To this aim we shall need some information concerning the density topology and its category analogue.

Let us start with the measure case. We say that x_0 is a density point of a measurable set $A \subset \mathbb{R}$ if and only if $\lim_{h\to 0^+} \frac{\lambda(A\cap(x_0-h,x_0+h))}{2h} = 1$ and a point of dispersion if and only if $\lim_{h\to 0^+} \frac{\lambda(A\cap(x_0-h,x_0+h))}{2h} = 0$. The classical theorem of Lebesgue (Lebesgue Density Theorem) says that $\lambda(A\triangle\Phi(A)) = 0$ for each measurable set $A \subset \mathbb{R}$, where $\Phi(A) = \{x \in \mathbb{R}:\ x \text{ is a density point of } A\}$. In other words almost all point of A are density points of A, almost all points of $\mathbb{R}\setminus A$ are density points of $\mathbb{R}\setminus A$, so almost all points of $\mathbb{R}\setminus A$ are points of dispersion of A. Denjoy in [Den15] has defined a class of approximately continuous real-valued functions of a real variable in the following way: f is approximately continuous at $x_0 \in \mathbb{R}$ if and only if there exists a measurable set $A \subset \mathbb{R}$ such that $x_0 \in \Phi(A)$ and $\lim\limits_{x\to x_0, x\in A} f(x) = f(x_0)$. For many years approximate continuity was studied without the connection with any topology. The density topology was defined by Haupt and Pauc in [HP52], but the first real interest in this topology appeared when Goffman and Waterman, and Goffman, Neugebauer and Nishiura have published important papers [GW61, GNN61]. For more information see [Oxt80]. The density topology on the real line is the family $\mathcal{T}_d = \{A \in \mathcal{L}:\ A \subset \Phi(A)\}$. Observe that $\mathcal{T}_d \subset \mathcal{L}$ and $\mathcal{T}_d$ is closed under arbitrary unions while $\mathcal{L}$ is closed only under countable unions. The family $\mathcal{U}(x_0)$ of neighbourhoods of x_0 in $\mathcal{T}_d$ consists, as usual, of all sets $U \in \mathcal{T}_d$ such that $x_0 \in U$. A function $f\colon \mathbb{R} \to \mathbb{R}$ is approximately continuous at x_0 if and only if for each $\varepsilon > 0$ there exists $U \in \mathcal{U}(x_0)$ such that $f(U) \subset (f(x_0) - \varepsilon, f(x_0) + \varepsilon)$. It can be proved that these two definitions of approximate continuity coincide (see, for example [Wil02]). A function $f\colon \mathbb{R} \to \mathbb{R}$ is approximately continuous if and only if for each set O open in the natural topology $\mathcal{T}$ on $\mathbb{R}$ $f^{-1}(O) \in \mathcal{T}_d$. We shall consider the restriction of $\mathcal{T}_d$ to $[0, 1]$.

The following theorem will be used in the sequel.

Theorem 4.12 *If a function $f\colon [0, 1] \to \mathbb{R}$ is measurable, then it is approximately continuous almost everywhere.*

Proof Let $\varepsilon > 0$. By virtue of Lusin's theorem there exists a closed set $F \subset [0, 1]$ such that $\lambda([0, 1] \setminus F) < \varepsilon$ and a function $f|_F$ is continuous. From the Lebesgue Density Theorem it follows that $\lambda(F \cap \Phi(F)) = \lambda(F)$ and according to the Denjoy definition f is approximately continuous at each point of $F \cap \Phi(F)$, so it is approximately continuous except perhaps on the set of measure less than ε. From the arbitrariness of ε it follows that f is approximately continuous almost everywhere. □

Now we are ready to consider pointwise definition of unique upper and lower measurable boundaries. Suppose that $f\colon [0, 1] \to [-1, 1]$ is an arbitrary function.

Definition 4.13 We shall say that a function $M\colon [0, 1] \to [-1, 1]$ defined by: $M(x_0) = \inf_{U\in\mathcal{U}(x_0)} (\sup\{f(x)\colon x \in U\})$ is a measurable maximum of f at x_0. The function $m\colon [0, 1] \to [-1, 1]$ defined in the following way: $m(x_0) = \sup_{U\in\mathcal{U}(x_0)} (\inf\{f(x)\colon x \in U\})$ is called a measurable minimum of f at x_0.

Theorem 4.14 *M is an upper measurable boundary and m is a lower measurable boundary of f.*

Proof We shall show that $M = G$ (see Theorem 4.7) almost everywhere. Suppose that $\{x \in [0,1]: M(x) < G(x)\} \notin \mathcal{N}$. There exists $\varepsilon > 0$ such that $A = \{x \in [0,1]: M(x) + \varepsilon < G(x)\} \notin \mathcal{N}$. G is measurable, so it is approximately continuous almost everywhere, so there exists $x_0 \in A$ at which G is approximately continuous. Let $U_1 \in \mathcal{U}(x_0)$ be such that $G(x) > G(x_0) - \frac{\varepsilon}{2}$ for $x \in U_1$ and let $U_2 \in \mathcal{U}(x_0)$ be such that $\sup\{f(x): x \in U_2\} < M(x_0) + \frac{\varepsilon}{2}$. For $x \in U_1 \cap U_2$ we have $f(x) < M(x_0) + \frac{\varepsilon}{2} < G(x_0) - \frac{\varepsilon}{2}$. If we put

$$h(x) = \begin{cases} G(x) & \text{for } x \notin U_1 \cap U_2 \\ M(x_0) + \frac{\varepsilon}{2} & \text{for } x \in U_1 \cap U_2 \end{cases}$$

then $f \leqslant h$ and it is not true that $G \leqslant h$—a contradiction. Hence $G(x) \leqslant M(x)$.

Suppose now that $\{x \in [0,1]: G(x) < M(x)\} \notin \mathcal{N}$. Then there exists $\varepsilon > 0$ such that $A = \{x \in [0,1]: G(x) + \varepsilon < M(x)\} \notin \mathcal{N}$. Let $x_0 \in A$ be a point of approximate continuity of G. There exists $U_1 \in \mathcal{U}(x_0)$ such that $G(x) < G(x_0) + \frac{\varepsilon}{2}$ for $x \in U_1$. Put $U_2 = (U_1 \setminus \{x \in [0,1]: f(x) > G(x)\}) \cup \{x_0\}$. Observe that $U_2 \in \mathcal{U}(x_0)$, since $\{x \in [0,1]: f(x) > G(x)\} \in \mathcal{N}$. If $x \in U_2$, then $f(x) \le G(x) < g(x_0) + \frac{\varepsilon}{2} < M(x_0) - \frac{\varepsilon}{2}$, so $\sup\{f(x): x \in U_2\} \le M(x_0) - \frac{\varepsilon}{2}$—a contradiction. Finally $M = G$ almost everywhere and obviously M is measurable. The proof for m is analogous. □

To consider the categorial maximum and minimum of an arbitrary function we shall need basic information concerning the category analogue of the density topology. It was introduced in [PWW85], see also [CLO91] or [Wil02].

A point 0 is a category density point of a set $A \subset \mathbb{R}$ having the Baire property if and only if the sequence $\{\chi_{(nA)\cap[-1,1]}\}_{n\in\mathbb{N}}$ of characteristic functions converges to $\chi_{[-1,1]}$ in category. This definition is based upon the observation that 0 is a density point of $A \subset \mathbb{R}$ if and only if $\{\chi_{(nA)\cap[-1,1]}\}_{n\in\mathbb{N}}$ of characteristic functions converges in measure to $\chi_{[-1,1]}$. A point x_0 is a category density point of $A \subset \mathbb{R}$ if and only if 0 is a category density point of $A - x_0$ (here $nA = \{nx: x \in A\}$ and $A - x_0 = \{x - x_0: x \in A\}$. A point x_0 is called a category dispersion point of A if and only if it is a category density point of $\mathbb{R} \setminus A$. It means that $\{\chi_{(nA)\cap[-1,1]}\}_{n\in\mathbb{N}}$converges in category to $\chi_{\varnothing}$. If $\Phi_c(A) = \{x \in \mathbb{R}: x \text{ is a category density point of } A\}$, where $A \in \mathcal{B}$, then $A \triangle \Phi_c(A) \in \mathcal{K}$. The proof is essentially simpler than that of Lebesgue Density Theorem due to the fact that the structure of set having the Baire property is simpler than of measurable set. The family $\mathcal{T}_c = \{A \in \mathcal{B}: A \subset \Phi_c(A)\}$ is a topology on the real line called the category density topology. If $x_0 \in [0,1]$, then $\mathcal{U}(x_0)$ will mean now the family of all neighbourhoods of x_0 in a category density topology (reduced to $[0,1]$). A function $f: [0,1] \to \mathbb{R}$ is $\mathcal{K}$-approximately continuous at $x_0 \in [0,1]$ if and only if for each $\varepsilon > 0$ there exists $U \in \mathcal{U}(x_0)$ such that $f(U) \subset (f(x_0) - \varepsilon, f(x_0) + \varepsilon)$. A function $f: [0,1] \to \mathbb{R}$ is $\mathcal{K}$-approximately continuous if and only if for each set O open in the natural topology on $\mathbb{R}$ $f^{-1}(O) \in \mathcal{T}_c$.

Theorem 4.15 *If $f : [0, 1] \to \mathbb{R}$ has the Baire property, then it is $\mathcal{K}$-approximately continuous $\mathcal{K}$-almost everywhere.*

Proof There exists a residual set $A \subset [0, 1]$ such that the restriction $f|_A$ is continuous. The set A has the Baire property, so $A = O \triangle P$, where O is open and $P \in \mathcal{K}$. If $x_0 \in O \setminus P$, then $x_0 \in \Phi_c(A)$, so f is $\mathcal{K}$-approximately continuous at x_0 and obviously $O \setminus P$ is residual in $[0, 1]$. □

Suppose that $f : [0, 1] \to [-1, 1]$ is arbitrary function.

Definition 4.16 We shall say that a function $M_c : [0, 1] \to [-1, 1]$ defined in the following way: $M_c(x_0) = \inf_{U \in \mathcal{U}(x_0)} (\sup\{f(x) : x \in U\})$, is a category maximum of f at x_0. The function $m_c : [0, 1] \to [-1, 1]$ defined by $m_c(x_0) = \sup_{U \in \mathcal{U}(x_0)} (\inf\{f(x) : x \in U\})$ is called a category minimum of f at x_0.

Theorem 4.17 *M_c is an upper categorial boundary of f and m_c is a lower categorial boundary of f, both M_c and m_c have the Baire property.*

Proof The proof that $M_c = G_c$ and $m_c = g_c$ $\mathcal{K}$-almost everywhere is essentially the same as in measure case. □

Now we shall compare functions M and m with upper and lower measurable boundaries considered by Blumberg [Blu35]. Suppose that $f : [0, 1] \to [-1, 1]$ is an arbitrary function. Let $B_1(y)$ and $B_2(y)$ be measurable covers of $\{x \in [0, 1] : f(x) > y\}$ and $\{x \in [0, 1] : f(x) < y\}$, respectively.

Definition 4.18 We shall say that a function $u : [0, 1] \to [-1, 1]$ defined in the following way: $u(x_0) = \inf \{y : x_0$ is a dispersion point of $B_1(y)\}$ is called upper measurable (B)-boundary of f at x_0, the function $l : [0, 1] \to [-1, 1]$ defined by $l(x_0) = \sup \{y : x_0$ is a dispersion point of $B_2(y)\}$ is called lower measurable (B)-boundary of f at x_0,

Theorem 4.19 *For arbitrary function $f : [0, 1] \to [-1, 1]$ $u(x) = M(x)$ and $l(x) = m(x)$ everywhere.*

Proof If $y > u(x_0)$, then x_0 is a point of dispersion of $B_1(y)$. Hence $U = ([0, 1] \setminus B_1(y)) \cup \{x_0\} \in \mathcal{U}(x_0)$ and $\sup\{f(x) : x \in U\} \le y$, so from the arbitrariness of $y > u(x_0)$ it follows that $M(x_0) \le u(x_0)$. If $y < u(x_0)$, then x_0 is not a point of dispersion of $B_1(y)$. We shall show that if $U \in \mathcal{U}(x_0)$, then $U \cap \{x \in [0, 1] : f(x) > y\} \ne \emptyset$. Suppose that this is not the case. Then $[0, 1] \setminus U$ is a measurable cover of $\{x \in [0, 1] : f(x) > y\}$ and x_0 is a dispersion point of $[0, 1] \setminus U$, a contradiction. So $\sup\{f(x) : x \in U\} > y$. From arbitrariness of $y < u(x_0)$ it follows that $M(x_0) \ge u(x_0)$. The proof for l is analogous. □

Blumberg measurable boundaries of an arbitrary function have the following analogues for category: Suppose again that $f : [0, 1] \to [-1, 1]$, $B_1(y)$ and $B_2(y)$ are Baire covers of $\{x \in [0, 1] : f(x) > y\}$ and $\{x \in [0, 1] : f(x) < y\}$, respectively.

Definition 4.20 We shall say that a function $u_c\colon [0,1] \to [-1,1]$ defined in the following way: $u_c(x_0) = \inf\{y\colon x_0 \text{ is a category dispersion point of } B_1(y)\}$ is called upper categorial (B)-boundary of f at x_0, the function $l_c\colon [0,1] \to [-1,1]$ defined by $l_c(x_0) = \sup\{y\colon x_0 \text{ is a category dispersion point of } B_2(y)\}$ is called lower categorial (B)-boundary of f at x_0.

Theorem 4.21 *For arbitrary function* $f\colon [0,1] \to [-1,1]$ $u_c(x) = M_c(x)$ *and* $l_c(x) = m_c(x)$ *everywhere.*

Proof The proof is essentially the same as in measure case. □

Goffman and Waterman in [GW60] have considered upper and lower limits in measure of a sequence of measurable functions. Their proofs used the machinery of Lebesgue measure and integral. Wagner in [Wag81] has proved analogous result for sequences of functions having the Baire property. Her proof used only basic properties of the pair $(\mathcal{B}, \mathcal{K})$ including the countable chain condition and works also in the case of measure. In the sequel we shall present results from [Wag81] assuming that $(\mathcal{S}, \mathcal{I})$ is $(\mathcal{L}, \mathcal{N})$ or $(\mathcal{B}, \mathcal{K})$. All considered real-valued functions will be defined on $[0, 1]$.

Definition 4.22 If $\{f_n\}_{n\in\mathbb{N}}$ and $\{g_n\}_{n\in\mathbb{N}}$ are sequences of $\mathcal{S}$-measurable functions defined on $[0, 1]$, then we say that these sequences are equivalent if and only if $\{f_n - g_n\}_{n\in\mathbb{N}}$ converges to zero with respect to $\mathcal{I}$ (in measure or in category).

It is clear that the relation defined above is an equivalence relation. Let $\mathcal{F}$ be the equivalence class including $\{f_n\}_{n\in\mathbb{N}}$.

Definition 4.23 We shall say that an $\mathcal{S}$-measurable function U (resp. L) is an upper (lower) limit of a sequence $\{f_n\}_{n\in\mathbb{N}}$ with respect to $\mathcal{I}$ if and only if

$$U = \inf\left[\limsup_n g_n : \{g_n\}_{n\in\mathbb{N}} \in \mathcal{F}\right]$$
$$\left(L = \sup\left[\liminf_n g_n : \{g_n\}_{n\in\mathbb{N}} \in \mathcal{F}\right]\right).$$

The existence and measurability of U and L follows from Theorem 4.4.

Lemma 4.24 *If U and L are upper and lower limits with respect to $\mathcal{I}$ of a sequence* $\{f_n\}_{n\in\mathbb{N}}$, *then* $L \le U$.

Proof Suppose that $U(x) < L(x)$ on a set $B \notin \mathcal{I}$. Then there exists a sequence $\{h_n\}_{n\in\mathbb{N}} \in \mathcal{F}$ such that $\limsup_n h_n(x) < L(x)$ on the set $C \notin \mathcal{I}$ and there exists a sequence $\{g_n\}_{n\in\mathbb{N}} \in \mathcal{F}$ such that $\limsup_n h_n(x) < \liminf_n g_n(x)$ on the set $D \notin \mathcal{I}$. We have

$$D = \bigcup_{w\in\mathbb{Q}\wedge w>0} \{x : \limsup_n h_n(x) + w < \liminf_n g_n(x)\},$$

where $\mathbb{Q}$ is the set of all rational numbers, and so there exists a number w_0 such that the set $\{x : \limsup_n h_n(x) + w_0 < \liminf_n g_n(x)\}$ does not belong to $\mathcal{I}$. Obviously

$$\{x : \limsup_n h_n(x) + w_0 < \liminf_n g_n(x)\}$$
$$= \bigcup_{n=1}^{\infty} \{x : \sup(h_n(x), h_{n+1}(x), \ldots) + w_0 < \inf(g_n(x), g_{n+1}(x), \ldots)\},$$

hence there exists a positive integer n_0 such that the set

$$E = \{x : \sup(h_{n_0}(x), h_{n_0+1}(x), \ldots) + w_0 < \inf(g_{n_0}(x), g_{n_0+1}(x), \ldots)\}$$

does not belong to $\mathcal{I}$. Therefore for each $x \in E$ and for each $n \geq n_0$ we have $g_n(x) - h_n(x) > w_0 > 0$, and so a sequence $\{g_n\}_{n \in \mathbb{N}}$ is not equivalent to $\{h_n\}_{n \in \mathbb{N}}$—a contradiction. □

Lemma 4.25 *Let $\{f_n\}_{n \in \mathbb{N}}$ be a sequence of $\mathcal{S}$-measurable functions. If $\{f_n\}_{n \in \mathbb{N}}$ does not converge to zero with respect to $\mathcal{I}$, then at least one of the following conditions is fulfilled:*

1. *there exist a subsequence $\{f_{m_n}\}_{n \in \mathbb{N}}$ of $\{f_n\}_{n \in \mathbb{N}}$, a set $A_0 \in \mathcal{S} \setminus \mathcal{I}$ and a positive integer k_0 such that for each subsequence $\{f_{m_{p_n}}\}_{n \in \mathbb{N}}$ of $\{f_{m_n}\}_{n \in \mathbb{N}}$ we have $\limsup_n f_{m_{p_n}}(x) \geq 1/k_0$ $\mathcal{I}$-a.e. on A_0;*
2. *there exist a subsequence $\{f_{m_n}\}_{n \in \mathbb{N}}$ of $\{f_n\}_{n \in \mathbb{N}}$, a set $A_0 \in \mathcal{S} \setminus \mathcal{I}$ and a positive integer k_0 such that for each subsequence $\{f_{m_{p_n}}\}_{n \in \mathbb{N}}$ of $\{f_{m_n}\}_{n \in \mathbb{N}}$ we have $\liminf_n f_{m_{p_n}}(x) \leq -1/k_0$ $\mathcal{I}$-a.e. on A_0.*

Proof Suppose on the contrary that for each subsequence $\{f_{m_n}\}_{n \in \mathbb{N}}$ of $\{f_n\}_{n \in \mathbb{N}}$, for each set $A \in \mathcal{S} \setminus \mathcal{I}$ and for each natural number k there exists a subsequence $\{f_{m_{p_n}}\}_{n \in \mathbb{N}}$ of $\{f_{m_n}\}_{n \in \mathbb{N}}$ such that $\limsup_n f_{m_{p_n}}(x) < 1/k$ on the set $A' \subset A$ such that $A' \in \mathcal{S} \setminus \mathcal{I}$ and that for each subsequence $\{f_{m_n}\}_{n \in \mathbb{N}}$ of $\{f_n\}_{n \in \mathbb{N}}$, for each set $A \in \mathcal{S} \setminus \mathcal{I}$ and for each natural number k there exists a subsequence $\{f_{m_{p_n}}\}_{n \in \mathbb{N}}$ of $\{f_{m_n}\}_{n \in \mathbb{N}}$ such that $\liminf_n f_{m_{p_n}}(x) > -1/k$ on the set $A'' \subset A$ such that $A'' \in \mathcal{S} \setminus \mathcal{I}$.

As a first step we shall prove that under this supposition for each subsequence $\{f_{m_n}\}_{n \in \mathbb{N}}$ there exists a subsequence $\{\hat{f}_n^{(1)}\}_{n \in \mathbb{N}}$ of $\{f_{m_n}\}_{n \in \mathbb{N}}$ such that $\limsup_n \hat{f}_n^{(1)}(x) < 1$ $\mathcal{I}$-a.e. on $[0, 1]$.

Let $\{f_{m_n}\}_{n \in \mathbb{N}}$ be an arbitrary subsequence of $\{f_n\}_{n \in \mathbb{N}}$. Put $A = [0, 1]$ and $k = 1$. There exists a subsequence $\{f_n^{(1)}\}_{n \in \mathbb{N}}$ of $\{f_{m_n}\}_{n \in \mathbb{N}}$ such that

$$B_1 = \{x : \limsup_n f_n^{(1)}(x) < 1\} \notin \mathcal{I}.$$

Observe that $B_1 \in \mathcal{S}$.

Suppose that for each ordinal number $\alpha < \eta$, where $\eta < \Omega$, we have chosen a subsequence $\{f_n^{(\alpha)}\}_{n \in \mathbb{N}}$ of $\{f_{m_n}\}_{n \in \mathbb{N}}$ and set $B_\alpha \in \mathcal{S}$ such that if $\alpha_1 < \alpha_2 < \eta$, then $\{f_n^{(\alpha_2)}\}_{n \in \mathbb{N}}$ is almost a subsequence of $\{f_n^{(\alpha_1)}\}_{n \in \mathbb{N}}$, the sequence of sets $\{B_\alpha\}_{\alpha < \eta}$ is nondecreasing and $\limsup_n f_n^{(\alpha)}(x) < 1$ for $x \in B_\alpha$. If η has a predecessor and $[0, 1] \setminus B_{\eta-1} \in \mathcal{I}$, then we put $f_n^{(\eta)} = f_n^{(\eta-1)}$ for $n = 1, 2, \ldots$. If $[0, 1] \setminus B_{\eta-1} \notin \mathcal{I}$ then by virtue of the supposition we can choose a subsequence $\{f_n^{(\eta)}\}_{n \in \mathbb{N}}$ of

$\{f_n^{(\eta-1)}\}_{n\in\mathbb{N}}$ such that $\limsup_n f_n^{(\eta)}(x) < 1$ for $x \in B_\eta$ and $B_\eta \setminus B_{\eta-1} \notin \mathcal{I}$. Indeed, it suffices to take $A = [0, 1] \setminus B_{\eta-1}$ and $k = 1$.

If η does not have a predecessor and there exists an ordinal number $\alpha < \eta$ such that $[0, 1] \setminus B_\alpha \in \mathcal{I}$, then we put $f_n^{(\eta)} = f_n^{(\alpha)}$ for $n = 1, 2, \ldots$. If $[0, 1] \setminus B_\alpha \notin \mathcal{I}$ for $\alpha < \eta$, then we choose an increasing sequence $\{\alpha_k\}_{k\in\mathbb{N}}$ of ordinal numbers tending to η. Put $B_\eta = \bigcup_{\alpha<\eta} B_\alpha$. Then of course $B_\eta = \bigcup_{k=1}^{\infty} B_{\alpha_k}$. Let $\{f_n^{(\eta)}\}_{n\in\mathbb{N}}$ be a diagonal subsequence chosen from the family $\{f_n^{(\alpha_k)}\}_{n\in\mathbb{N}}$, $k = 1, 2 \ldots$. Then $\limsup_n f_n^{(\eta)}(x) < 1$ for $x \in B_\eta$.

We shall show that there exists an ordinal number $\beta < \Omega$ such that $[0, 1] \setminus B_\beta \in \mathcal{I}$. Suppose that $[0, 1] \setminus B_\alpha \notin \mathcal{I}$ for each $\alpha < \Omega$. From the definition of sets $\{B_\alpha\}_{\alpha<\Omega}$ it follows that if $\eta < \Omega$ has a predecessor, then $B_\eta \setminus B_{\eta-1} \notin \mathcal{I}$. Put $C_1 = B_1$, $C_\alpha = B_\alpha \setminus B_{\alpha-1}$, when $1 < \alpha < \Omega$ is an ordinal number having a predecessor. Then $C_\alpha \notin \mathcal{I}$ and $C_\alpha \cap C_\eta = \varnothing$ if $\alpha \neq \eta$, $\alpha, \eta < \Omega$, α and η are ordinal numbers having predecessors. We have thus obtained a contradiction with the countable chain condition. Hence there exists an ordinal number $\beta < \Omega$ such that $[0, 1] \setminus B_\beta \in \mathcal{I}$. If we put $\hat{f}_n^{(1)} = f_n^{(\beta)}$ for $n \in \mathbb{N}$, we obtain a subsequence for which $\limsup_n \hat{f}_n^{(1)}(x) < 1$ $\mathcal{I}$-a.e. on $[0, 1]$.

Repeating this argument for $k = 2$ and for $\{\hat{f}_n^{(1)}\}_{n\in\mathbb{N}}$, we obtain a subsequence $\{\hat{f}_n^{(2)}\}_{n\in\mathbb{N}}$ of $\{\hat{f}_n^{(1)}\}_{n\in\mathbb{N}}$ such that $\limsup_n \hat{f}_n^{(2)}(x) < \frac{1}{2}$ $\mathcal{I}$-a.e. on $[0, 1]$. Proceeding similarly for $k = 3, 4, \ldots$ and then choosing a diagonal subsequence $\{\overline{f}_n\}_{n\in\mathbb{N}}$, one can prove without difficulty that $\limsup_n \overline{f}_n(x) \le 0$ $\mathcal{I}$-a.e. on $[0, 1]$.

Using the second part of the supposition, we can choose a subsequence $\{f_{m_{p_n}}\}_{n\in\mathbb{N}}$ of $\{\overline{f}_n\}_{n\in\mathbb{N}}$ such that $\liminf_n f_{m_{p_n}}(x) \ge 0$ $\mathcal{I}$-a.e. on $[0, 1]$.

Hence from an arbitrary subsequence $\{f_{m_n}\}_{n\in\mathbb{N}}$ of $\{f_n\}_{n\in\mathbb{N}}$ we can choose a subsequence $\{f_{m_{p_n}}\}_{n\in\mathbb{N}}$ converging to zero $\mathcal{I}$ -a.e. on $[0, 1]$—a contradiction. □

Theorem 4.26 *Let f and f_n, $n \in \mathbb{N}$, be $\mathcal{S}$-measurable functions. A sequence $\{f_n\}_{n\in\mathbb{N}}$ converges with respect to $\mathcal{I}$ to a function f if and only if $U = L$, and then $f = U = L$.*

Proof Suppose that $\{f_n\}_{n\in\mathbb{N}}$ converges to f with respect to $\mathcal{I}$. Then a sequence $\{f_n\}_{n\in\mathbb{N}}$ is equivalent to a sequence $\{g_n\}_{n\in\mathbb{N}}$, where $g_n = f$ for $n = 1, 2, \ldots$. Hence $U \le f$, because $\limsup_n g_n = f$. Similarly one can prove that $L \ge f$. So $f \le L \le U \le f$ and $U = L$.

Suppose now that $U = L$. Without loss of generality we can suppose that $U \equiv 0$. Assume that $\{f_n\}_{n\in\mathbb{N}}$ does not converge to zero with respect to $\mathcal{I}$ and that the first condition of Lemma 4.24 is fulfilled (in the remaining case the proof is similar).

Let $\{g_n\}_{n\in\mathbb{N}}$ be a sequence equivalent to $\{f_n\}_{n\in\mathbb{N}}$. The $g_n = f_n + h_n$, $n \in \mathbb{N}$, where $\{h_n\}_{n\in\mathbb{N}}$ is a sequence of $\mathcal{S}$—measurable functions converging to zero with respect to $\mathcal{I}$. Let $\{f_{m_n}\}_{n\in\mathbb{N}}$ be a subsequence, A_0—a set and k_0—a positive integer described in

Lemma 4.24. There exists a subsequence $\{h_{m_{p_n}}\}_{n\in\mathbb{N}}$ of $\{h_{m_n}\}_{n\in\mathbb{N}}$ convergent to zero at each $x \in [0, 1] \setminus C$, where $C \in \mathcal{I}$. Let $x \in A_0 \setminus C$. Then

$$\limsup_n g_n(x) \geq \limsup_n g_{m_{p_n}}(x) = \limsup_n (f_{m_{p_n}}(x) + h_{m_{p_n}}(x)) \geq \frac{1}{k_0}.$$

From the arbitrariness of $\{g_n\}_{n\in\mathbb{N}}$ and from the definition of U we have $U(x) \geq \frac{1}{k_0}$ $\mathcal{I}$—a.e. on A_0. This contradiction ends the proof. □

Chapter 5
Continuity $\mathcal{I}$-a.e and Convergence $\mathcal{I}$-a.e

If $\mathcal{F}$ is a family of real-valued functions defined on [0, 1], then there exists a smallest family $\mathcal{B}(\mathcal{F})$ of real-valued functions defined on [0, 1] which contains $\mathcal{F}$ and is closed with respect to taking pointwise limits of sequences from $\mathcal{B}(\mathcal{F})$ (so called Baire system generated by $\mathcal{F}$). Let $\mathcal{F}_0 = \mathcal{F}$ and let $\mathcal{F}_\alpha$ be the family of all pointwise limits of sequences from $\bigcup_{\gamma<\alpha} \mathcal{F}_\gamma$ for each ordinal α. Then $\mathcal{F}_{\omega_1}$ is the Baire system generated by $\mathcal{F}$. We shall say that the ordinal α is the Baire order of $\mathcal{F}$ if α is the smallest ordinal such that $\mathcal{F}_\alpha = \mathcal{F}_{\alpha+1}$.

It is well known that the Baire order of the family of real-valued continuous functions defined on [0, 1] is ω_1 [Leb05]. Kantorovitch [Kan30] has shown that if $\mathcal{F}$ is a family of real-valued functions defined on [0, 1], which are continuous almost everywhere, then the Baire order of $\mathcal{F}$ is at least 2. This result was improved by R. D. Mauldin, who proved in [Mau73a] that this order is at least 3 and in [Mau73b] that the order of the family $\mathcal{F}$ is ω_1. If $\mathcal{F}$ is a family of real-valued functions defined on [0, 1] which are continuous $\mathcal{K}$-almost everywhere, then the Baire order of $\mathcal{F}$ equals to 1. This is a result of Kuratowski [Kur24], see also [Hau62].

Below we shall present the results of Kantorovitch completed by the observation of Z. Zalcwasser and the theorem of Kuratowski. We shall show also what happens when we consider continuous functions in the place of a.e. or $\mathcal{K}$-a.e. continuous functions and the convergence almost everywhere or $\mathcal{K}$-almost everywhere instead of pointwise convergence.

In the following theorem we shall need basic information concerning the classification of Young [You16]. Recall that the function $f : [0, 1] \to \mathbb{R}$ belongs to the first upper (lower) class of Young ($f \in u_1$, $f \in l_1$) if and only if it is a pointwise limit of a non-decreasing (non-increasing) sequence of continuous functions. That means that f is lower (upper) semicontinuous (semicontinuous from below, from above). The function f belongs to the second upper (lower) class of Young ($f \in u_2$, $f \in l_2$) if and only if it is a pointwise limit of a non-decreasing (resp. non-increasing) sequence of functions from l_1 (resp. u_1).

W. Wilczyński, *Convergence in Measure and in Category*, Lecture Notes in Mathematics 2394, https://doi.org/10.1007/978-3-032-23143-7_5

Observe that if $f \in u_2$, then $\{x \in [0, 1]\colon f(x) > a\} \in \mathcal{F}_\sigma$ for each $a \in \mathbb{R}$. Indeed, $f = \lim_{n\to\infty} f_n$, where $\{f_n\}_{n\in\mathbb{N}}$ is a non-decreasing sequence of upper semicontinuous functions. Hence $\{x \in [0, 1]\colon f(x) > a\} = \bigcup_{k=1}^{\infty} \bigcup_{n=1}^{\infty} \{x\colon f_n(x) \geq a + \frac{1}{k}\}$. But $\{x\colon f_n(x) \geq a + \frac{1}{k}\}$ is closed for each $n, k \in \mathbb{N}$, so the result follows.

Theorem 5.1 *([Kan30]) Suppose that $\{f_n\}_{n\in\mathbb{N}}$ is a sequence of real-valued functions defined on $[0, 1]$ such that f_n is continuous almost everywhere for each $n \in \mathbb{N}$ and $F(x) = \lim_{n\to\infty} f_n$ for each $x \in [0, 1]$. Then there exist two functions g and G defined on $[0, 1]$ such that $g \in l_2$, $G \in u_2$ and $g(x) \leq F(x) \leq G(x)$ for each $x \in [0, 1]$ and $g(x) = F(x) = G(x)$ almost everywhere.*

Proof Put $\overline{G}_{n,k}(x) = \min(f_n(x), f_{n+1}(x), \ldots, f_{n+k}(x))$ and $\overline{G}_n = \lim_{k\to\infty} \overline{G}_{n,k}$. Obviously $\overline{G}_{n,k}$ is continuous a.e. for each $n \in \mathbb{N}$ and $k \in \mathbb{N} \cup \{0\}$. The function $\overline{G}_n$ is upper semicontinuous almost everywhere, namely for each point of the set $\bigcap_{k=0}^{\infty} C_{n+k}$, where C_k is the set of points of continuity of f_k. We have also $\overline{G}_{n+1}(x) \geq \overline{G}_n$ for each $n \in \mathbb{N}$ and $x \in [0, 1]$, and $F(x) = \lim_{n\to\infty} \overline{G}_n(x)$ for each $x \in [0, 1]$. Put $G_n(x) = \limsup_{t\to x} \overline{G}_n(t)$ for $n \in \mathbb{N}$ and $x \in [0, 1]\big(G_n(x) = \lim_{\delta\to 0^+} \left(\sup\left\{\overline{G}_n(t)\colon t \in (x-\delta, x+\delta)\right\}\right)\big)$. Obviously $G_n(x) \geq \overline{G}_n(x)$ and $G_{n+1}(x) \geq G_n(x)$. Also each G_n is upper semicontinuous everywhere on $[0, 1]$ and $G_n(x) = \overline{G}_n(x)$ almost everywhere, i.e. at points at which $\overline{G}_n$ is upper semicontinuous. Now put $G(x) = \lim_{n\to\infty} G_n(x)$ for $x \in [0, 1]$. From the construction it follows that G belongs to u_2, $F(x) \leq G(x)$ everywhere, and $G(x) = F(x)$ almost everywhere.

The construction of g is self-explaining. □

Theorem 5.2 *([Zal23]) There exists measurable function $F\colon [0, 1] \to \mathbb{R}$ which is not a pointwise limit of functions continuous almost everywhere.*

Proof Let $E \subset [0, 1]$ be a residual set of type $\mathcal{G}_\delta$ such that $\lambda(E) = 0$ and put $F = \chi_E$. Suppose that there exists a sequence $\{f_n\}_{n\in\mathbb{N}}$ of functions continuous almost everywhere such that $\lim_{n\to\infty} f_n(x) = F(x)$ for $x \in [0, 1]$. Then by virtue of previous theorem there exists a function $G \in u_2$ such that $F(x) \leq G(x)$ and $F(x) = G(x)$ almost everywhere. The set $A = \{x \in [0, 1]\colon G(x) > \frac{1}{2}\}$ is of type $\mathcal{F}_\sigma$. Obviously $A \supset E$, so A is of second category and, as $\mathcal{F}_\sigma$ set, includes some interval. This contradicts the fact that $\lambda(E) = 0$. □

To prove the next theorem we shall need the classical theorem of Lusin.

Theorem 5.3 *([BBT97]) If $f\colon [0, 1] \to \mathbb{R}$ is measurable, then for each $\varepsilon > 0$ there exists a closed set $F \subset [0, 1]$ such that $\lambda([0, 1] \setminus F) < \varepsilon$ and $f|_F$ is continuous.*

Proof Suppose additionally that f is bounded, say $f\colon [0, 1] \to [0, 1]$. Fix $\varepsilon > 0$. Put $E_{n,m} = \{x \in [0, 1]\colon \frac{m-1}{n} \leq f(x) < \frac{m}{n}\}$ for $n \in \mathbb{N}$, $m \in \{1, 2, \ldots, n\}$ and let $f_n(x) = \frac{m-1}{n}$ for $x \in E_{n,m}$. A sequence $\{f_n\}_{n\in\mathbb{N}}$ converges uniformly to f on $[0, 1]$. For each $n \in \mathbb{N}$ and $m \in \{1, 2, \ldots, n\}$ there exists a closed set $F_{n,m} \subset E_{n,m}$ such that $\lambda(E_{n,m} \setminus F_{n,m}) < \frac{\varepsilon}{n \cdot 2^n}$ for each $m \in \{1, 2, \ldots, n\}$. For each $n \in \mathbb{N}$ the set $F_n =$

$\bigcup_{m=1}^{n} F_{n,m}$ is closed and the function $f_n|_{F_n}$ is continuous. Let $F = \bigcap_{n=1}^{\infty} \bigcup_{m=1}^{n} F_{n,m}$. The set F is closed, $\lambda([0,1] \setminus F) < \varepsilon$ and $\{f_n|_F\}$ consists of continuous functions. Hence $f|_F$ is continuous.

If f is unbounded, consider $g(x) = \frac{1}{2}\left(\frac{f(x)}{1+f(x)} + 1\right)$. □

Theorem 5.4 *If a function* $f: [0,1] \to \mathbb{R}$ *is measurable, then there exists a sequence* $\{f_n\}_{n\in\mathbb{N}}$ *of continuous functions such that* $\{f_n\}_{n\in\mathbb{N}}$ *converges to* f *a.e.*

Proof For each $n \in \mathbb{N}$ by virtue of previous theorem there exists a closed set $F_n \subset [0,1]$ such that $\lambda([0,1] \setminus F_n) < \frac{1}{n}$ and a function $f|_{F_n}$ is continuous. Put $H_n = \bigcup_{i=1}^{n} F_i$ for $n \in \mathbb{N}$. Then $f|_{H_n}$ is continuous and $\lambda([0,1] \setminus H_n) < \frac{1}{n}$ for each $n \in \mathbb{N}$. Let $f_n : [0,1] \to \mathbb{R}$ be a continuous function such that $f_n|_{H_n} = f|_{H_n}$ (the existence of such function is assured by Tietze's theorem). Then $\{f_n\}_{n\in\mathbb{N}}$ fulfills all requirements. □

Theorem 5.5 *If* $f: [0,1] \to \mathbb{R}$ *is measurable, then there exists a function* $g: [0,1] \to \mathbb{R}$ *in the second Baire class such that* $f = g$ *a.e.*

Proof Put $g = \limsup_{n\to\infty} f_n$, where $\{f_n\}_{n\in\mathbb{N}}$ is a sequence of functions described in previous theorem. Then $f = g$ a.e. and g is in the second Baire class. □

Remark 5.6 There exists a measurable function $f: [0,1] \to \mathbb{R}$ such that $\lambda(\{x \in [0,1]: f(x) \neq g(x)\}) > 0$ for each function g from the first Baire class.

Let $A \subset [0,1]$ be a measurable set such that $\lambda((a,b) \cap A) > 0$ and $\lambda((a,b) \setminus A) > 0$ for each interval $(a,b) \subset [0,1]$. Put $f = \chi_A$. If $g = f$ a.e., then for each interval $(a,b) \subset [0,1]$ we have $\lambda((a,b) \cap \{x: g(x) = 1\}) > 0$ and $\lambda((a,b) \cap \{x: g(x) = 0\}) > 0$, so both sets $\{x: g(x) = 1\}$ and $\{x: g(x) = 0\}$ are dense and g is discontinuous everywhere, so g is not in the first Baire class. □

Now we shall study the possibility of representation of functions having the Baire property as a limit of continuous or $\mathcal{K}$-a.e. continuous functions.

Theorem 5.7 ([Hau62], p. 323) *If* $f: [0,1] \to \mathbb{R}$ *has the Baire property, then there exists a sequence* $\{f_n\}_{n\in\mathbb{N}}$ *of real-valued functions defined on* $[0,1]$ *such that each* f_n *is continuous* $\mathcal{K}$*-almost everywhere and* $\lim_{n\to\infty} f_n(x) = f(x)$ *for each* $x \in [0,1]$.

Proof From the assumption if follows that there exists a residual set $C \subset [0,1]$ such that $f|_C$ is continuous. We can suppose that $C \in \mathcal{G}_\delta$. Then $[0,1] \setminus C$ is of the first category and $[0,1] \setminus C \in \mathcal{F}_\sigma$, so $[0,1] \setminus C = \bigcup_{k=1}^{\infty} F_k$, where F_k's are closed and nowhere dense. Moreover, we can and shall suppose that a sequence $\{F_k\}_{k\in\mathbb{N}}$ is ascending. Then $[0,1] \setminus C = F_1 \cup (F_2 \setminus F_1) \cup (F_3 \setminus F_2) \cup \ldots$ is a union of disjoint summands. If $x \in [0,1] \setminus C$, then there exists a unique number $k(x)$ such that $x \in F_{k(x)} \setminus F_{k(x)+1}$ and if a sequence $\{x_n\}_{n\in\mathbb{N}}$ of points from $[0,1] \setminus C$ converges to $x_0 \in C$, then $\lim_{n\to\infty} k(x_n) = \infty$.

Now we shall define a function $g: [0,1] \to \mathbb{R}$ in a following way: $g(x) = f(x)$ for $x \in C$ and if $x \in [0,1] \setminus C$, then we choose a point $z(x) \in C$ such that $|x - z| \leq \operatorname{dist}(x, C) + (k(x))^{-1}$ (where, as usual $\operatorname{dist}(x, C) = \inf\{|x - t|: t \in C\}$) and put

$g(x) = f(z(x))$. Observe that $g\colon [0,1] \to \mathbb{R}$ is continuous at each point of C, so it is continuous $\mathcal{K}$-almost everywhere. Indeed, let $x_0 \in C$ and $\{x_n\}_{n\in\mathbb{N}}$ be a sequence such that $x_n \in [0,1]$ for each $n \in \mathbb{N}$ and $\lim_{n\to\infty} x_n = x_0$. We may restrict our considerations to the two cases that the terms of a sequence belong either to C or to $[0,1] \setminus C$. In the first case we have $g(x_n) = f(x_n)$, $g(x_0) = f(x_0)$ and $\lim_{n\to\infty} g(x_n) = g(x_0)$ since $f|_C$ is continuous. In the second case $\lim_{n\to\infty} \operatorname{dist}(x_n, C) = 0$ and $\lim_{n\to\infty} k(x_n) = \infty$. Hence $\lim_{n\to\infty} |x_n - z(x_n)| = 0$ so $\lim_{n\to\infty} z(x_n) = x_0$ and $\lim_{n\to\infty} g(x_n) = \lim_{n\to\infty} f(z(x_n)) = f(x_0)$ again because of the continuity of $f|_C$.

Put

$$f_k(x) = \begin{cases} f(x) & \text{for } x \in C \cup F_k \\ g(x) & \text{for } x \in [0,1] \setminus (C \cup F_k) \end{cases}.$$

Fix $k \in \mathbb{N}$. We shall show that f_k is continuous at each point of C. Let $x_0 \in C$ and $\{x_n\}_{n\in\mathbb{N}}$ is a sequence convergent to x_0. As before, if $\{x_n\}_{n\in\mathbb{N}}$ consists of points from C, the conclusion is immediate. If all x_n's are in $[0,1] \setminus C$, then there exists n_0 such that for $n \geq n_0$ we have $x_n \notin F_k$, so $f_k(x_n) = f(x_n)$ and so $\lim_{n\to\infty} f_k(x_n) = \lim_{n\to\infty} f(x_n) = f(x_0) = f_k(x_0)$. Hence $\{f_k\}_{k\in\mathbb{N}}$ consists of functions continuous $\mathcal{K}$-almost everywhere.

At the same time $f_k(x) = f(x)$ for $x \in C$ and each $k \in \mathbb{N}$ and also $f_k(x) = f(x)$ for $k \geq k(x)$ if $x \in [0,1] \setminus C = \bigcup_{k=1}^{\infty} F_k$, so $\lim_{k\to\infty} f_k(x) = f(x)$ for $x \in [0,1]$. □

We shall prove that in the theorem above the continuity $\mathcal{K}$-a.e. can be replaced by the continuity everywhere and simultaneous change of the convergence everywhere onto the convergence $\mathcal{K}$-almost everywhere. To this aim we shall use the theorem of Sierpiński [Sie30], more precisely the part of this theorem [the original result is valid for each Baire class α, the extended function is in Baire class $(\alpha + 1)$]. We shall follow the reasoning of Sierpiński.

Theorem 5.8 *If $E \subset [0,1]$ is an arbitrary set and $f\colon E \to \mathbb{R}$ is continuous, then there exists a function $\varphi\colon [0,1] \to \mathbb{R}$ in Baire class 1 such that $\varphi|_E = f$.*

Proof Let $g(x) = \max(f(x), 0)$ and $h(x) = \max(-f(x), 0)$ for $x \in E$. Obviously $f = g - h$ and we shall prove the thesis for g, the proof for h is similar and the difference of two Baire one functions is Baire one.

For each $a \in \mathbb{R}$ the set $\{x \in E\colon g(x) < a\}$ is of the form $E \cap G$, where G is open, since g is continuous.

Let $\{x \in E\colon g(x) < \frac{k}{2^n}\} = E \cap G_n^k$, $k \in \mathbb{N} \cup \{0\}$, $n \in \mathbb{N}$.

We shall need the lemma. □

Lemma 5.9 *There exists a double sequence of open sets $\{S_n^k\}_{n\in\mathbb{N}, k\in\mathbb{N}\cup\{0\}}$ such that $S_n^k \subset S_n^{k+1}$ and $E \cap S_n^k = E \cap G_n^k$ for each $n, k \in \mathbb{N}$.*

Proof of the Lemma Put $S_n^0 = \varnothing$ for $n \in \mathbb{N}$ and $S_1^k = G_1^1 \cup G_1^2 \cup \ldots \cup G_1^k$ for $k \in \mathbb{N}$. We shall define by induction

$$S_n^{2k-1} = \left(G_n^{2k-1} \cap S_{n-1}^k\right) \cup S_{n-1}^{k-1}, \quad S_n^{2k} = S_{n-1}^k \quad \text{for } k \in \mathbb{N}.$$

The inclusion $S_1^k \subset S_1^{k+1}$ is obvious. Suppose that $S_n^k \subset S_n^{k+1}$ for each k. We have

$$S_{n+1}^{2k-1} = \left(G_{n+1}^{2k-1} \cap S_n^k\right) \cup S_n^{k-1} \subset S_n^k \cup S_n^{k-1} = S_n^k = S_{n+1}^{2k}$$

and

$$S_{n+1}^{2k} = S_n^k \subset \left(G_{n+1}^{2k+1} \cap S_n^{k+1}\right) \cup S_n^k = S_{n+1}^{2k+1},$$

so finally $S_{n+1}^k \subset S_{n+1}^{k+1}$ for $k \in \mathbb{N}$, which has to be proved.

From the definition of G_n^k we have

$$E \cap G_{n+1}^{2k} = E \cap G_n^k \quad \text{and} \quad E \cap G_n^{k-1} \subset E \cap G_{n+1}^{2k-1} \subset E \cap G_n^k,$$

so $\bigcup_{i=1}^k (E \cap G_1^i) = E \cap G_1^k$, so $E \cap S_1^k = E \cap G_1^k$ for $k \in \mathbb{N}$.

Suppose now that $E \cap S_n^k = E \cap G_n^k$ for each $k \in \mathbb{N}$. Then

$$E \cap S_{n+1}^{2k} = E \cap S_n^k = E \cap G_n^k = E \cap G_{n+1}^{2k}$$

and

$$\begin{aligned} E \cap S_{n+1}^{2k-1} &= \left(E \cap G_{n+1}^{2k-1} \cap S_n^k\right) \cup (E \cap S_n^{k-1}) = \left(E \cap G_{n+1}^{2k-1} \cap G_n^k\right) \cup (E \cap G_n^{k-1}) \\ &= E \cap G_{n+1}^{2k-1}, \end{aligned}$$

so $E \cap S_{n+1}^k = E \cap G_n^k$ for each $k \in \mathbb{N}$, which has to be proved. □

Let's return to the proof of the theorem. Put $\varphi_n(x) = \frac{k}{2^n}$ if $x \in S_n^{k+1} \setminus S_n^k$ for $k \in \mathbb{N} \cup \{0\}$, $n \in \mathbb{N}$, and $\varphi_n(x) = 0$ if $x \notin \bigcup_{k=1}^{\infty} S_n^k$ for $n \in \mathbb{N}$. The sets $S_n^{k+1} \setminus S_n^k$, $k, n \in \mathbb{N}$, as well as $[0, 1] \setminus \bigcup_{k=1}^{\infty} S_n^k$, $n \in \mathbb{N}$, are at the same time of type $\mathcal{G}_\delta$ and $\mathcal{F}_\sigma$, so all φ_n's are Baire one.

We shall show that the sequence $\{\varphi_n\}_{n \in \mathbb{N}}$ of functions is uniformly convergent. It is enough to show that $|\varphi_n(x) - \varphi_{n-1}(x)| \leq \frac{1}{2^n}$ for $n \in \mathbb{N}$ and $x \in [0, 1]$.

If $x \in S_n^{k+1} \setminus S_n^k$, then $\varphi_n(x) = \frac{k}{2^n}$ and if $k = 2l$, then $S_n^{k+1} = S_n^{2l+1} \subset S_{n-1}^{l+1} \cup S_{n-1}^l = S_{n-1}^{l+1}$ and $S_n^{2l} = S_{n-1}^l$, so $S_n^{2l+1} \setminus S_n^{2l} \subset S_{n-1}^{l+1} \setminus S_{n-1}^l$, so from the definition of φ_n it follows that $\varphi_{n-1}(x) = \frac{l}{2^{n-1}}$, hence $\varphi_n(x) = \varphi_{n-1}(x)$.

Now if $x \in S_n^{k+1} \setminus S_n^k$ and $k = 2l - 1$, then $S_n^{2l} = S_{n-1}^l$ and $S_n^{2l-1} \supset S_{n-1}^{l-1}$, so $S_n^{2l} \setminus S_n^{2l-1} \subset S_{n-1}^l \setminus S_{n-1}^{l-1}$, hence $\varphi_{n-1}(x) = \frac{l-1}{2^{n-1}}$ and finally $\varphi_n(x) - \varphi_{n-1}(x) = \frac{1}{2^n}$.

If $x \notin \bigcup_{k=1}^{\infty} S_n^k$, then $\varphi_n(x) = 0$. Since $S_{n-1}^k \subset S_n^{2k+1}$ for $k \in \mathbb{N}$, we have $\bigcup_{n=1}^{\infty} S_{n-1}^k \subset \bigcup_{k=1}^{\infty} S_n^k$ and $x \notin \bigcup_{n=1}^{\infty} S_{n-1}^k$, so $\varphi_{n-1}(x) = 0$ and $\varphi_n(x) = \varphi_{n-1}(x)$. Hence the sequence $\{\varphi_n\}_{n \in \mathbb{N}}$ is uniformly convergent and the function $\varphi = \lim_{n \to \infty} \varphi_n$ is Baire one.

We shall show that $\varphi(x) = g(x)$ for $x \in E$. Indeed, if $x \in E$, then for each $n \in \mathbb{N}$ there exists $k \geq 0$ such that

$$\frac{k}{2^n} \leq g(x) < \frac{k+1}{2^n},$$

so $x \in E \cap (G_n^{k+1} \setminus G_n^k) = E \cap (S_n^{k+1} \setminus S_n^k)$, so $\varphi_n(x) = \frac{k}{2^n}$ and $0 \leq g(x) - \varphi_n(x) < \frac{1}{2^n}$ for each $n \in \mathbb{N}$ and the results follows. □

Theorem 5.10 *If* $F\colon [0,1] \to \mathbb{R}$ *has the Baire property, then there exists a sequence* $\{f_n\}_{n\in\mathbb{N}}$ *of continuous functions defined on* $[0,1]$ *such that* $F = \lim_{n\to\infty} f_n$ *$\mathcal{K}$-almost everywhere. In other words there exists a function* $g \in \mathcal{B}_1$ *such that* $F = g$ *$\mathcal{K}$-a.e.*

Proof Since F has the Baire property, there exists a residual set $E \subset [0,1]$ such that $F|_E$ is continuous. From the previous theorem it follows that there exists Baire one function $\varphi\colon [0,1] \to \mathbb{R}$ such that $\varphi|_E = F|_E$, hence there exists a sequence $\{f_n\}_{n\in\mathbb{N}}$ of continuous functions defined on $[0,1]$ which converges pointwise to φ, so $\{f_n\}_{n\in\mathbb{N}}$ converges $\mathcal{K}$-a.e. to F. □

It is not difficult to observe that Theorems inverse to 5.7 and 5.10 are also true.

Now we shall present theorem of Saks and Sierpiński [SS28], which concerns arbitrary real-valued functions defined on [0, 1].

Suppose that $(\mathcal{S}, \mathcal{I})$ is a pair of σ-algebra and σ-ideal of subsets of $[0,1]$ such that $\mathcal{I} \subset \mathcal{S}$. Let $A, E \subset [0,1]$. We shall say that A is $\mathcal{S}$-cover of E if and only if $A \in \mathcal{S}$ $A \supset E$ and for each $B \in \mathcal{S}$ if $B \supset E$, then $A \setminus B \in \mathcal{I}$. This condition is equivalent to the following: if $C \subset A \setminus E$ and $C \in \mathcal{S}$, then $C \in \mathcal{I}$. Indeed, suppose that the first condition is fulfilled. Take $C \subset A \setminus E$, $C \in \mathcal{S}$ and suppose that $C \notin \mathcal{I}$. Since $A \setminus C \in \mathcal{S}$ and $A \setminus C \supset E$, we have $A \setminus (A \setminus C) = C \notin \mathcal{I}$ - a contradiction. So the second condition holds. If the first condition is not fulfilled, which means that there exists $B \in \mathcal{S}$ such that $B \supset E$ and $A \setminus B \notin \mathcal{I}$, then the second condition does not hold, because $A \setminus B \in \mathcal{S}$ and $A \setminus B \subset A \setminus E$.

Let

$$S^* = \{A \subset [0,1]\colon \text{ for each } B \subset [0,1] \text{ if } B \supset A \text{ and } B \in \mathcal{S}, \text{ then } [0,1] \setminus B \in \mathcal{I}\}.$$

Observe that $\mathcal{L}^*$ is the family of subsets of $[0,1]$ for which the outer Lebesgue measure equals 1. The sets belonging to $\mathcal{B}^*$ shall be called outer residual.

In the proof of the next theorem we shall need the following lemmas.

Lemma 5.11 *If* $A_1 \in S^*$ *and* $[0,1] \setminus A_2 \in \mathcal{I}$, *then* $A_1 \cap A_2 \in S^*$.

Proof In the proof we shall use interchangeably the denotation A' and $[0,1] \setminus A$. Take $B \supset A_1 \cap A_2$, $B \in \mathcal{S}$. We shall show that $[0,1] \setminus B \in \mathcal{I}$. Observe first that $A_1 \subset B \cup (A_1 \cap A_2')$. Indeed, $A_1 = (A_1 \cap A_2) \cup (A_1 \cap A_2') \subset B \cup (A_1 \cap A_2')$. Moreover, $B \cup (A_1 \cap A_2') \in \mathcal{S}$, since $A_1 \cap A_2' \in \mathcal{I}$. Hence $[0,1] \setminus \big(B \cup (A_1 \cap A_2')\big) \in \mathcal{I}$. Simultaneously $[0,1] \setminus B \subset \big([0,1] \setminus \big(B \cup (A_1 \cap A_2')\big)\big) \cup A_2'$, because

$$\begin{aligned}[0,1] \setminus B = [0,1] \cap B' &= \big([0,1] \cap B' \cap (A_1 \cap A_2')'\big) \cup \big([0,1] \cap B' \cap (A_1 \cap A_2')\big) \\ &\subset \big([0,1] \setminus \big(B \cup (A_1 \cap A_2')\big)\big) \cup A_2'.\end{aligned}$$

Hence $[0,1] \setminus B \in \mathcal{I}$ as required, because both summands belong to $\mathcal{I}$. □

Lemma 5.12 *If $E \subset [0, 1]$ and $g\colon E \to \mathbb{R}$ is a function such that* $\operatorname{card}(g(E)) \le \aleph_0$, *then there exist $\mathcal{S}$-measurable function $\varphi\colon [0, 1] \to \mathbb{R}$ and a set $H \subset E$ such that $H^* = E^*$ and $g(x) = \varphi(x)$ for $x \in H$.*

Proof Let $g(E) = \{a_1, a_2, \dots, a_n, \dots\}$ and $E_n = \{x \in E\colon g(x) = a_n\}$ for $n \in \mathbb{N}$. Obviously $E = \bigcup_{n=1}^{\infty} E_n$. We shall define the sequence $\{M_n\}_{n\in\mathbb{N}}$ of sets belonging to $\mathcal{S}$ in the following way: Let $M_1 = E_1^*$. If $M_1, \dots, M_{n-1}$ have been defined, then let $M_n = (E_n \setminus \bigcup_{i=1}^{n-1} M_i)^*$. Taking, if necessary, $M_n \setminus \bigcup_{i=1}^{n-1} M_i$ we can (and shall) suppose that all M_n's are disjoint. Put $H = \bigcup_{n=1}^{\infty}(M_n \cap E_n)$ and let $M = H^*$. We have $E_n \subset \bigcup_{i=1}^{n} M_i$ for $n \in \mathbb{N}$. Hence $E_n \setminus \bigcup_{i=1}^{n-1} M_i \subset M_n \cap E_n$, so $M \supset E_n \setminus \bigcup_{i=1}^{n-1} M_i$ for $n \in \mathbb{N}$. Since $M_n = (E_n \setminus \bigcup_{i=1}^{n-1} M_i)^*$, $M_n \setminus M \in \mathcal{I}$ for $n \in \mathbb{N}$. Simultaneously $\bigcup_{n=1}^{\infty} M_n \setminus M = \bigcup_{n=1}^{\infty}(M_n \setminus M)$, so $\bigcup_{n=1}^{\infty} M_n \setminus M \in \mathcal{I}$ and $E = \bigcup_{n=1}^{\infty} E_n \subset \big(\bigcup_{n=1}^{\infty} M_n \cap M\big) \cup \big(\big(\bigcup_{n=1}^{\infty} M_n\big) \setminus M\big)$, from which it follows that $E^* \Subset M = H^*$ (where $\Subset$ means that $E^* \setminus M \in \mathcal{I}$). Since $H \subset E$, we have also $H^* \Subset E^*$. Modifying, if necessary, H^*, i.e. choosing $\mathcal{S}$-covers differing from H^* and E^* on a set belonging to $\mathcal{I}$ we obtain $H^* = E^*$.

Put now $\varphi(x) = a_n$ for $x \in M_n$, $n \in \mathbb{N}$ and $\varphi(x) = 0$ for $x \in [0, 1] \setminus \bigcup_{n=1}^{\infty} M_n$. The function φ is $\mathcal{S}$-measurable and $g(x) = \varphi(x)$ for $x \in M_n \cap E_n, n \in \mathbb{N}$, so $g(x) = \varphi(x)$ for $x \in H$. □

Lemma 5.13 *If $E \subset [0, 1]$ and $f\colon E \to \mathbb{R}$ is an arbitrary function, then for each $\varepsilon > 0$ there exist an $\mathcal{S}$-measurable function $\varphi\colon [0, 1] \to \mathbb{R}$ and a set $H \subset E$ such that $H^* = E^*$ and $|f(x) - \varphi(x)| < \varepsilon$ for $x \in H$.*

Proof Put $g(x) = \varepsilon\big[\frac{f(x)}{\varepsilon}\big]$ for $x \in E$ ($[t]$ stands for the whole part of t). Obviously we have $0 \le f(x) - g(x) < \varepsilon$ for each $x \in E$. Since $\operatorname{card}(g(\varepsilon)) \le \aleph_0$, by virtue of Lemma 5.12 there exist an $\mathcal{S}$-measurable function $\varphi\colon [0, 1] \to \mathbb{R}$ and a set $H \subset E$ such that $H^* = E^*$ and $g(x) = \varphi(x)$ for $x \in H$. □

Theorem 5.14 *([SS28]) Suppose that $(\mathcal{S}, \mathcal{I})$ is a pair such that each set $E \subset [0, 1]$ has $\mathcal{S}$-cover. Then for each real-valued function f defined on $[0, 1]$ there exists $\mathcal{S}$-measurable function $\varphi\colon [0, 1] \to \mathbb{R}$ such that for each $\varepsilon > 0$ $\{x \in [0, 1]\colon |f(x) - \varphi(x)| < \varepsilon\} \in \mathcal{S}^*$.*

Proof First we shall prove that the theorem holds when the range of f is countable. We shall write E^* for an $\mathcal{S}$-cover of $E \subset [0, 1]$. Let f be an arbitrary real-valued function defined on $[0, 1]$. Put $E = [0, 1]$, $\varepsilon = \frac{1}{2}$ and apply Lemma 5.13. We shall denote the resulting function and set by φ_1 and H_1. Obviously $[0, 1] \Subset H_1^*$.

Suppose that we have defined φ_{n-1} and H_{n-1}. Let φ_n and H_n be the function and the set obtained from Lemma 5.13 for $E = H_{n-1}$ and $\varepsilon = 2^{-n}$. So we obtained a sequence $\{\varphi_n\}_{n\in\mathbb{N}}$ of $\mathcal{S}$-measurable functions and $\{H_n\}_{n\in\mathbb{N}}$ such that $[0, 1] \Subset H_n^*$ and $H_{n+1} \subset H_n \subset [0, 1]$ for $n \in \mathbb{N}$. We have $|f(x) - \varphi_n(x)| < \frac{1}{2^n}$ for $x \in H_n$ and $|f(x) - \varphi_{n+1}(x)| < \frac{1}{2^{n+1}}$ for $x \in H_{n+1}$, so $|\varphi_{n+1}(x) - \varphi_n(x)| < \frac{1}{2^{n-1}}$ for $x \in H_{n+1}$. The set $P_n = \{x \in [0, 1]\colon |\varphi_{n+1}(x) - \varphi_n(x)| < \frac{1}{2^{n-1}}\}$ is $\mathcal{S}$-measurable, since all φ_n's are $\mathcal{S}$-measurable. Moreover, $P_n \supset H_{n+1}$, so

$[0, 1] \Subset P_n$ for each $n \in \mathbb{N}$. Hence $[0, 1] \Subset \bigcap_{i=1}^{\infty} P_i = P \in \mathcal{S}^*$. From the definition of P_n we have $|\varphi_{n+1}(x) - \varphi_n(x)| < \frac{1}{2^{n-1}}$ for $x \in P_n$, $n \in \mathbb{N}$. Hence $|\varphi_{n+1}(x) - \varphi_n(x)| < \frac{1}{2^{n-1}}$ for $x \in P$ and $n \in \mathbb{N}$. That means that the sequence $\{\varphi_n\}_{n\in\mathbb{N}}$ is uniformly convergent on P. Put $\varphi(x) = \lim_{n\to\infty} \varphi_n(x)$ for $x \in P$ and $\varphi(x) = 0$ elsewhere. The function $\varphi\colon [0, 1] \to \mathbb{R}$ is $\mathcal{S}$-measurable and $|\varphi(x) - \varphi_n(x)| < \frac{1}{2^{n-2}}$ for $x \in P$, so $|f(x) - \varphi(x)| < \frac{1}{2^{n-3}}$ for $x \in P \cap H_n$. From Lemma 5.12 it follows that $P \cap H_n \in \mathcal{S}^*$. This ends the proof. □

Corollary 5.15 *For each real-valued function f defined on $[0, 1]$ there exists a function φ from the second Baire class such that*

$$\lambda^* \left(\{x \in [0, 1]\colon |f(x) - \varphi(x)| < \varepsilon\}\right) = 1$$

for each $\varepsilon > 0$.

For each real-valued function f defined on $[0, 1]$ there exists a function φ from the first Baire class such that $\{x \in [0, 1]\colon |f(x) - \varphi(x)| < \varepsilon\}$ is outer residual for each $\varepsilon > 0$.

Chapter 6
Atomic Spaces

If $(X, \mathcal{S}, \mu)$ is a finite measure space, then a set $A \in \mathcal{S}$ is called an atom if $\mu(A) > 0$ and if, moreover, for each $B \subset A$, $B \in \mathcal{S}$, either $\mu(B) = 0$ or $\mu(A) = \mu(B)$. If a set $A \subset \mathcal{S}$ contains no atom, then A and the measure μ are called atomless. It is easy to see that if $A \in \mathcal{S}$ is an atom of μ and $\mu(A \triangle B) = 0$, then B is also an atom of μ. Observe also that for each finite measure space $(X, \mathcal{S}, \mu)$ there exists a decomposition of X into disjoint sets: $X = \bigcup_{n=0}^{\infty} X_n$, where X_0 is either empty or an atomless set and each of the sets $X_1, X_2, \ldots$ is either empty or an atom. If X_0 is empty, then μ is called atomic measure.

Theorem 6.1 ([Mar55]) *If $(X, \mathcal{S}, \mu)$ is a finite measure space and μ is an atomic measure, then a sequence $\{f_n\}_{n \in \mathbb{N}}$ of real-valued $\mathcal{S}$-measurable functions defined on X converges almost everywhere to f if and only if it converges to f in measure.*

Proof Since the convergence a.e. yields the convergence in measure, we shall prove only the converse implication. To this aim we need the following lemma:

Lemma 6.2 *If A is an atom of μ and $f : X \to \mathbb{R}$ is $\mathcal{S}$-measurable function, then f is constant a.e. on A.*

Proof of the lemma Let $A_n = f^{-1}([-n, n]) \cap A$. The sequence $\{A_n\}_{n \in \mathbb{N}}$ of $\mathcal{S}$-measurable sets is ascending and $A = \bigcup_{n=1}^{\infty} A_n$. Hence $\mu(A) = \lim_{n \to \infty} \mu(A_n)$, so there exists $n_0 \in \mathbb{N}$ such that $\mu(A_{n_0}) = \mu(A)$. Consider $f^{-1}([-n_0, 0]) \cap A$, $f^{-1}([0, n_0]) \cap A$. If $\mu\left(f^{-1}([-n_0, 0]) \cap A\right) = \mu(A_{n_0}) = \mu(A)$, put $I_1 = [-n_0, 0]$. If $\mu\left(f^{-1}([-n_0, 0]) \cap A\right) = 0$, put $I_1 = [0, n_0]$. Suppose that we have chosen $I_1, I_2, \ldots, I_n$. Consider $J_0 = [\inf I_n, \frac{1}{2}(\inf I_n + \sup I_n)]$ and $J_1 = [\frac{1}{2}(\inf I_n + \sup I_n), \sup I_n]$. If $\mu(f^{-1}(J_0) \cap A) = \mu(A)$, put $I_{n+1} = J_0$, if not, put $I_{n+1} = J_1$. So we have constructed a descending sequence $\{I_n\}_{n \in \mathbb{N}}$ such that $\mu(f^{-1}(I_n) \cap A) = \mu(A)$ for each $n \in \mathbb{N}$ and $\bigcap_{n=1}^{\infty} I_n = \{y_0\}$, $\bigcap_{n=1}^{\infty} \left(f^{-1}(I_n)\right) \cap A = f^{-1}(y_0) \cap A$ and $\mu\left(f^{-1}(y_0) \cap A\right) = \mu(A)$. □

Proof of theorem Suppose now that a sequence $\{f_n\}_{n \in \mathbb{N}}$ converges in measure to f_0 and A is an atom. Then for each $n \in \mathbb{N} \cup \{0\}$ there exists $A_n \subset A$ such that

W. Wilczyński, *Convergence in Measure and in Category*, Lecture Notes in Mathematics 2394, https://doi.org/10.1007/978-3-032-23143-7_6

$\mu(A_n) = \mu(A)$ and f_n is constant on A_n, say $f_n(x) = c_n$. Put $B = \bigcap_{n=1}^{\infty} A_n$. We have $\mu(B) = \mu(A)$, so B is an atom of μ and $f_n(x) = c_n$ for $x \in B$ and for each $n \in \mathbb{N} \cup \{0\}$. Let $\varepsilon > 0$. From the assumption it follows that there exists $n_0 \in \mathbb{N}$ such that $\mu(\{x\colon |f_n(x) - f(x)| \geq \varepsilon\}) < \mu(A)$ for $n \geq n_0$. Hence $|c_n - c_0| < \varepsilon$ for $n \geq n_0$ and $\lim_{n\to\infty} c_n = c_0$, which means that $\{f_n\}_{n\in\mathbb{N}}$ converges to f almost everywhere on each atom A, so almost everywhere on X. □

Theorem 6.3 ([Mar55], see also [Gri66]) *If $(X, \mathcal{S}, \mu)$ is a finite measure space and μ is non atomic measure, then there exist a sequence $\{f_n\}_{n\in\mathbb{N}}$ of real-valued $\mathcal{S}$-measurable functions convergent to 0 in measure which is not convergent almost everywhere.*

Proof From the assumption it follows that there exists $X_0 \subset X$ such that $\mu(X_0) > 0$ and X_0 is atomless. According to [Hal65] (p. 174) for each number c with $0 < c < \mu(X_0)$ there exists a set $B \subset X_0$, $B \in \mathcal{S}$ such that $\mu(B) = c$.

Let B_0 and B_1 be sets such that $B_0 \in \mathcal{S}$, $B_1 \in \mathcal{S}$, $B_0 \cap B_1 = \emptyset$, $B_0 \cup B_1 = X_0$ and $\mu(B_0) = \mu(B_1) = \frac{1}{2}\mu(X_0)$. Suppose that we have defined sets B_{i_1}, $i_1 = 0, 1$; B_{i_1,i_2}, $i_1, i_2 = 0, 1$; ...; $B_{i_1,i_2,\ldots,i_n}$, $i_1, i_2, \ldots, i_n = 0, 1$. Then for each sequence $i_1, i_2, \ldots, i_n$ consisting of 0's and 1's there exist sets $B_{i_1,i_2,\ldots,i_n,0}$ and $B_{i_1,i_2,\ldots,i_n,1}$ belonging to $\mathcal{S}$, disjoint, such that $B_{i_1,i_2,\ldots,i_n,0} \cup B_{i_1,i_2,\ldots,i_n,1} = B_{i_1,i_2,\ldots,i_n}$ and $\mu(B_{i_1,i_2,\ldots,i_n,0}) = \mu(B_{i_1,i_2,\ldots,i_n,1}) = \frac{1}{2}\mu(B_{i_1,i_2,\ldots,i_n})$. If f_n is a characteristic function of the set on n-th place in the sequence $B_0, B_1, B_{0,1}, B_{1,0}, B_{1,1}, \ldots$, then the sequence $\{f_n\}_{n\in\mathbb{N}}$ converges in measure to 0 but diverges at each point of X_0. □

Suppose that $(X, \mathcal{S})$ is a measurable space and $\mathcal{I} \subset \mathcal{S}$ is a proper σ-ideal. Recall that the pair $(\mathcal{S}, \mathcal{I})$ fulfills the countable chain condition (ccc) if and only if each disjoint subfamily of $\mathcal{S} \setminus \mathcal{I}$ is at most denumerable. We shall say that the set $A \in \mathcal{S} \setminus \mathcal{I}$ is an atom if for each $B \subset A$ if $B \in \mathcal{S}$ then either $B \in \mathcal{I}$ or $A \setminus B \in \mathcal{I}$ and we shall say that the pair $(\mathcal{S}, \mathcal{I})$ is atomic if for each $C \in \mathcal{S} \setminus \mathcal{I}$ there exists an atom $A \subset C$. If $(\mathcal{S}, \mathcal{I})$ is atomic, then from (ccc) it follows immediately that $X = \bigcup_k A_k$ (the union is finite or denumerable), where the disjoint family $\{A_k\}_{k\in\mathbb{N}}$ consists of atoms. In another words if $A \sim B$ is an equivalent relation on $\mathcal{S}$ defined by $A \triangle B \in \mathcal{I}$, then the quotient Boolean algebra $\mathcal{S}/_{\mathcal{I}}$ is atomic.

Recall that a topological space is called a Baire space if the intersection of countably many open dense sets is dense. In [W-BW80] it is proved that if the Souslin hypothesis is true than the convergence $\mathcal{I}$-a.e. is equivalent to the convergence with respect to $\mathcal{I}$ if and only if the pair $(\mathcal{S}, \mathcal{I})$ is atomic. Miller in [Mil96] has proved that if Souslin hypothesis is false, then there exists a pair $(\mathcal{S}, \mathcal{I})$ which is not atomic and the convergence $\mathcal{I}$-a.e. is equivalent to the convergence with respect to $\mathcal{I}$. We shall present proofs of both theorems. To this aim we need basic facts concerning trees.

A tree is a partially ordered set $(T, <)$ such that for each $s \in T$ the set $\{t \in T : t < s\}$ is well ordered by $<$. The height of s in $(T, <)$ is the ordinal number of $\{t \in T : t < s\}$ and will be denoted ht(s). The αth level of T is the set $T_\alpha = \{s \in T : \mathrm{ht}(s) = \alpha\}$. The height of T (ht$T$) is the ordinal number $\min\{\alpha\colon T_\alpha = \emptyset\}$. We shall use also the following denotation: $T_{<\alpha} = \bigcup\{T_\beta\colon \beta < \alpha\}$.

A branch of tree T is a maximal chain of T. A path of T is any chain which is also an initial part of T, which means that if s belongs to a path and $t < s$, then t also belongs to this path. An α-branch is a branch having ordinal number α, similarly we define α-chain and α-path. An antichain in T is a set $A \subset T$ such that any two disjoint elements x, y of A are incomparable, which means that neither $x < y$ nor $y < x$.

A tree is normal (compare [Tod84]) if $\text{card}\{t \in T : \text{ht}(t) = 0\} = 1$; if $s, t \in T$ and $\text{ht}(t) = \text{ht}(s)$ is a limit ordinal and if $\{u \in T : u < s\} = \{u \in T : u < t\}$, then $s = t$; if t is not maximal in T, then there are at least two points at the next level (immediate successors of t). It is worth mentioning that there are non-equivalent definitions of a normal tree (see for example [Jec78]).

A Souslin tree is a tree T with $\text{ht}(T) = \omega_1$ and each branch and antichain is countable. Souslin hypothesis is equivalent to the non-existence of a Souslin tree.

Theorem 6.4 ([W-BW80]) *Assume Souslin hypothesis. If the pair $(\mathcal{S}, \mathcal{I})$ fulfills (ccc), then the convergence $\mathcal{I}$-a.e. of sequences of real-valued $\mathcal{S}$-measurable functions defined on X is equivalent to the convergence with respect to $\mathcal{I}$ if and only if the pair $(\mathcal{S}, \mathcal{I})$ is atomic.*

Proof Suppose that $(\mathcal{S}, \mathcal{I})$ is atomic. Let $\{f_n\}_{n\in\mathbb{N}}$ be the sequence convergent to f with respect to $\mathcal{I}$. We shall prove that $\{f_n\}_{n\in\mathbb{N}}$ converges to f $\mathcal{I}$-a.e (it is sufficient because convergence $\mathcal{I}$-a.e. implies the convergence with respect to $\mathcal{I}$). The idea of the proof is essentially the same as in [Gri66].

Let $\{A_k\}_{k\in\mathbb{N}}$ be the sequence of atoms for which $X = \bigcup_{k\in\mathbb{N}} A_k$. If $f : X \to \mathbb{R}$ is $\mathcal{S}$-measurable, then form Lemma 6.2 it follows that there exists a sequence $\{a^k\}_{k\in\mathbb{N}}$ of real numbers such that $f = \sum_{k\in\mathbb{N}} a^k \chi_{A_k}$ $\mathcal{I}$-a.e. Similarly for each $n \in \mathbb{N}$ there exists a sequence $\{a_n^k\}_{k\in\mathbb{N}}$ of real numbers such that $f_n = \sum_{k\in\mathbb{N}} a_n^k \chi_{A_k}$. From the assumption and from the definition of the convergence with respect to $\mathcal{I}$ it follows that for each $k \in \mathbb{N}$ and for each subsequence $\{a_{n_m}^k\}_{m\in\mathbb{N}}$ of $\{a_n^k\}_{k\in\mathbb{N}}$ there exists a subsequence $\{a_{n_{m_p}}^k\}_{p\in\mathbb{N}}$ convergent to a^k, hence $a^k = \lim_{n\to\infty} a_n^k$. It means that $f_n \underset{n\to\infty}{\to} f$ $\mathcal{I}$-a.e.

Suppose that $(\mathcal{S}, \mathcal{I})$ is not atomic. We shall construct a sequence $\{f_n\}_{n\in\mathbb{N}}$ of $\mathcal{S}$-measurable real-valued functions convergent to zero with respect to $\mathcal{I}$ which does not converge $\mathcal{I}$-a.e. to zero.

From the assumption if follows that there exists $A \in \mathcal{S} \setminus \mathcal{I}$ which does not include atoms, which means that for each $B \subset A$ if $B \in \mathcal{S} \setminus \mathcal{I}$, then there exists $B_1, B_2 \in \mathcal{S} \setminus \mathcal{I}$ such that $B_1 \cup B_2 = B$ and $B_1 \cap B_2 = \emptyset$. We shall call the pair $\{B_1, B_2\}$ a proper subdivision of B. We shall prove now that if $A \in \mathcal{S} \setminus \mathcal{I}$ is a set which does not include atoms, then there exists a sequence $\{d_n\}_{n\in\mathbb{N}}$ of subdivisions of A (by a subdivision we shall mean as usual a finite disjoint family of subset of A with the union equal to A) such that

(a) for each n d_{n+1} is a refinement of d_n;
(b) for each sequence $\{A_n\}_{n\in\mathbb{N}}$ of sets such that $A_n \in d_n$ we have $\bigcap_{n\in\mathbb{N}} A_n \in \mathcal{I}$;
(c) for each n $d_n \subset \mathcal{S} \setminus \mathcal{I}$.

Let $\{d_n^{(1)}\}_{n\in\mathbb{N}}$ be a sequence of subdivisions defined in the following way: $d_1^{(1)} = \{A_{1,1}^{(1)}, A_{1,2}^{(1)}\}$ is a proper subdivision of A and for every natural n let $d_{n+1}^{(1)} = \{A_{n+1,1}^{(1)}, ..., A_{n+1,2^n}^{(1)}\}$, where $\{A_{n+1,2k-1}^{(1)}, A_{n+1,2k}^{(1)}\}$ is a proper subdivision of $A_{n,k}^{(1)}$ for $k = 1, ..., 2^{n-1}$. It is not difficult to verify that $\{d_n^{(1)}\}_{n\in\mathbb{N}}$ has the properties a) and c). Denote $\mathcal{I}_1 = \{\bigcap_{n\in\mathbb{N}} B_n : B_n \in d_n^{(1)}\}$. If $\mathcal{I}_1 \subset \mathcal{I}$, then put $d_n = d_n^{(1)}$ for every natural n. If $\mathcal{I}_1 \setminus \mathcal{I} \neq \emptyset$, then we shall construct next sequences of subdivisions to obtain the desired result.

Suppose that for all ordinal numbers $\alpha < \eta$, where $\eta < \omega_1$ we have already defined the sequences $\{d_n^{(\alpha)}\}_{n\in\mathbb{N}}$ of subdivisions of A and related families $\mathcal{I}_\alpha = \{\bigcap_{n\in\mathbb{N}} B_n : B_n \in d_n^{(\alpha)}\}$ having the properties (a) and (c) and the following two properties for all ordinal numbers $\alpha, \beta < \eta, \alpha < \beta$:

(i) there exists a number N such that for every $n \geq N$ $d_n^{(\beta)}$ is a refinement of $d_n^{(\alpha)}$;
(ii) the family $\mathcal{I}_\beta$ is essentially finer than the family $\mathcal{I}_\alpha$ (it means that for every pair of sets C, D such that $C \in \mathcal{I}_\alpha \setminus \mathcal{I}$, $D \in \mathcal{I}_\beta \setminus \mathcal{I}$ and $D \subset C$ we have $C \setminus D \in \mathcal{S} \setminus \mathcal{I}$).

Observe also that each $\mathcal{I}_\alpha$, $\alpha < \eta$, is a disjoint countable family of sets. We shall define a sequence of subdivisions $\{d_n^{(\eta)}\}_{n\in\mathbb{N}}$. Consider two cases.

The first case. A number η has the predecessor $(\eta - 1)$. Suppose that $\mathcal{I}_{\eta-1} \setminus \mathcal{I} \neq \emptyset$, because in the contrary case the construction would be finished at earlier stage. From (ccc) it follows that $\mathcal{I}_{\eta-1} \setminus \mathcal{I} = \{D_1, D_2, ...\}$. Let for every natural n $\{D_n', D_n''\}$ be a proper subdivision of D_n. Let $d_1^{(\eta)}$ be a common refinement of $d_1^{(\eta-1)}$ and $\{A \setminus D_1, D_1', D_1''\}$. If we have defined $d_k^{(\eta)}$ for $k = 1, ..., n$, then let $d_{n+1}^{(\eta)}$ be a common refinement of $d_{n+1}^{(\eta-1)}$ and $\{A \setminus D_{n+1}, D_{n+1}', D_{n+1}''\}$. It is not difficult to verify that the sequence $\{d_n^{(\eta)}\}_{n\in\mathbb{N}}$ has the properties (a) and (c) and that now (i) and (ii) are fulfilled for $\alpha, \beta \leq \eta$.

The second case. The number η is a limit number. Suppose that for every $\alpha < \eta$ $\mathcal{I}_\alpha \setminus \mathcal{I} \neq \emptyset$, because in the contrary case the construction would be finished at earlier stage. Let $\{\alpha_k\}_{k\in\mathbb{N}}$ be an increasing sequence of ordinal numbers tending to η. From the property (i) it follows that for every k there exists N_k such that for each $n \geq N_k$ $d_n^{(\alpha_k+1)}$ is simultaneously a refinement of $d_n^{(\alpha_1)}, d_n^{(\alpha_2)}, ..., d_n^{(\alpha_k)}$. Obviously we can suppose that $\{N_k\}_{k\in\mathbb{N}}$ is an increasing sequence of natural numbers. Put $d_1^{(\eta)} = d_1^{(\alpha_1)}$ and for every $k \geq 2$ $d_k^{(\eta)} = d_{N_k-1}^{(\alpha_k)}$. It is not difficult to verify that the sequence $\{d_n^{(\eta)}\}_{n\in\mathbb{N}}$ has the properties (a) and (c) and that now (i) and (ii) are again fulfilled for $\alpha, \beta \leq \eta$.

Our aim is now to prove that for some $\eta < \omega_1$ we obtain $\mathcal{I}_\eta \subset \mathcal{I}$. Suppose the contrary. Then by transfinite induction we obtain a transfinite sequence $\{\{d_n^{(\eta)}\}_{n\in\mathbb{N}}\}_{\eta<\omega_1}$ of sequences of subdivisions such that for every $\alpha, \beta < \omega_1$ if $\alpha < \beta$, then $\mathcal{I}_\beta$ is essentially finer than $\mathcal{I}_\alpha$.

Put $T = \bigcup_{\eta<\omega_1}(\mathcal{I}_\alpha \setminus \mathcal{I})$. Consider the partial ordering $\prec$ defined in the following way: $A \prec B$ if and only if $A \supset B$ for $A, B \in T$. It is easy to prove that $(T, \prec)$ is a tree. From the assumption $\mathcal{I}_\eta \setminus \mathcal{I} \neq \emptyset$ for every $\eta < \omega_1$ we conclude that $\mathrm{card}(T) = \aleph_1$. We assume the Souslin's hypothesis, so $(T, \prec)$ is not a Souslin's tree. Hence it has a nondenumerable branch or a nondenumerable antichain. In the first case let $\{A_\alpha\}_{\alpha<\omega_1}$ be this branch. Then $\{A_\alpha \setminus A_{\alpha+1}\}_{\alpha<\omega_1}$ is a nondenumerable family of disjoint sets from $\mathcal{S} \setminus \mathcal{I}$—a contradiction with (ccc). In the second case let $\{B_i\}_{i\in\mathcal{I}}$ be an antichain in T such that $\text{card}(\mathcal{I}) = \aleph_1$. From the construction it follows that the sets belonging to the antichain are pairwise disjoint—again a contradiction with (ccc).

Hence there exists an ordinal number $\eta < \omega_1$ such that $\mathcal{I}_\eta \subset \mathcal{I}$. Put $d_n = d_n^{(\eta)}$ for every n. The sequence $\{d_n\}_{n\in\mathbb{N}}$ has obviously the properties (a), (b) and (c).

Let $\{B_n\}_{n\in\mathbb{N}}$ be a sequence of sets in which every set from every partition of the sequence $\{d_n\}_{n\in\mathbb{N}}$ arrives exactly one time. Put $f_n = \chi_{B_n}$ for every n. It is nearly obvious that for every $x \in A$ the sequence $\{f_n(x)\}_{n\in\mathbb{N}}$ is divergent, so the sequence of function $\{f_n\}_{n\in\mathbb{N}}$ does not converge $\mathcal{I}$-a.e. (because $A \notin \mathcal{I}$). We shall show that $\{f_n\}_{n\in\mathbb{N}}$ converges to zero with respect to $\mathcal{I}$.

Let $\{f_{n_m}\}_{m\in\mathbb{N}}$ be a subsequence of $\{f_n\}_{n\in\mathbb{N}}$. We can suppose (taking suitable subsequence of $\{f_{n_m}\}_{m\in\mathbb{N}}$ if necessary) that the corresponding sequence $\{B_{n_m}\}_{m\in\mathbb{N}}$ has the following property: $B_{n_m} \in d_{l_m}$, where $\{l_m\}_{m\in\mathbb{N}}$ is increasing sequence of positive integer. Two cases are possible:

1. there exists a subsequence $\{n_{m_p}\}_{p\in\mathbb{N}}$ of $\{n_m\}_{m\in\mathbb{N}}$ such that $\{B_{n_{m_p}}\}_{p\in\mathbb{N}}$ is descending,
2. there is no such sequence.

In the first case take a subsequence $\{f_{n_{m_p}}\}_{p\in\mathbb{N}}$. This subsequence converges $\mathcal{I}$-e.a. to zero (more precisely, except on $\bigcap_{p\in\mathbb{N}} B_{n_{m_p}}$, because from the condition b) it follows that $\bigcap_{p\in\mathbb{N}} B_{n_{m_p}} \in \mathcal{I}$.

In the second case we proceed as follows: let $\{B_{n_{m_1}}, ..., B_{n_{m_{p_1}}}\}$ be a maximal descending sequence of $\{B_{n_m}\}_{m\in\mathbb{N}}$ (in this case such a sequence does exist). We observe that each set in the sequence $\{B_{n_{m_{p_1+1}}}, B_{n_{m_{p_1+2}}}, ...\}$ is disjoint with $B_{n_{m_{p_1}}}$. For the set $B_{n_{m_{p_1+1}}}$ we also take a maximal descending sequence $\{B_{n_{m_{p_1}}}, ..., B_{n_{m_{p_2}}}\}$. Now each set in the sequence $\{B_{n_{m_{p_2+1}}}, B_{n_{m_{p_2+2}}}, ...\}$ is disjoint with $B_{n_{m_{p_1}}}$ and $B_{n_{m_{p_2}}}$. Proceeding further by induction we obtain a sequence $\{B_{n_{m_{p_i}}}\}_{i\in\mathbb{N}}$ of disjoint sets. The subsequence $\{f_{n_{m_{p_i}}}\}_{i\in\mathbb{N}}$ converges to zero at each point. So in both cases it is possible to find a subsequence $\{f_{n_{m_{p_i}}}\}_{i\in\mathbb{N}}$ convergent $\mathcal{I}$-e.a. to zero, which means that $\{f_n\}_{n\in\mathbb{N}}$ converges to zero with respect to $\mathcal{I}$. □

In this chapter meager means of the first category. We shall show that Souslin hypothesis is essential in the above theorem.

Theorem 6.5 ([Mil96]) *Suppose that there exists a Souslin tree (so Souslin hypothesis is false). Then there exists a topological Baire space X such that each family of disjoint open sets is countable, each nonempty open set in X is nonmeager, X has not isolated points and if $\mathcal{I}$ is the σ-ideal of meager subsets of X, then convergence $\mathcal{I}$-almost everywhere of a sequence of real-valued Baire measurable functions is the same as convergence in category.*

Proof Suppose that $(T, <)$ is a Souslin tree and X—the set of all branches of T. We shall construct a topology on X. For each $s \in T$ let $C_s = \{b \in X : s \in b\}$ and let $\{C_s : s \in T\}$ be an open basis for the topology in X. Observe that if $s, t \in T$ and

$s \le t$ (which means $s < t$ or $s = t$), then $C_s \cap C_t = C_t$, if s and t are incomparable, then $C_s \cap C_t = \emptyset$. If $A \subset T$ is a maximal antichain in T and $A_1, A_2 \subset A$ such that $A_1 \cup A_2 = A$ and $A_1 \cap A_2 = \emptyset$, then both sets $\bigcup_{s \in A_1} C_s$, $\bigcup_{s \in A_2} C_s$ are open and closed, since $X = \left(\bigcup_{s \in A_1} C_s\right) \cup \left(\bigcup_{s \in A_2} C_s\right)$.

In particular each C_s is clopen. The space X has no isolated points. Indeed, if $s \in T$, then C_s has at least two elements, because s has at least two incomparable immediate successors. Moreover X fulfills the countable chain condition. If $\{U_i\}_{i \in I}$ is a disjoint family of non-empty open sets, then for each $i \in I$ there exists $s_i \in T$ such that $C_{s_i} \subset U_i$. The set $\{s_i\}_{i \in I}$ consists of incomparable in T elements, so it is at most countable.

We shall need some lemmas:

Lemma 6.6 *The intersection of countably many open dense sets in X contains an open dense set, so X is a Baire space.*

Proof of the lemma Suppose $\{U_n\}_{n \in \mathbb{N}}$ is a sequence of open dense subsets of X. Let $A_n \subset T$ be an antichain maximal with respect to the property that $C_s \subset U_n$ for each $s \in A_n$. We shall show that A_n is also a maximal antichain in T. Suppose that it is not the case, so there exists $s_0 \in T$ which is incomparable with each $s \in A_n$. Since U_n is dense, $U_n \cap C_{s_0} \neq \emptyset$. For each $s \in A_n$ we have $C_s \cap C_{s_0} = \emptyset$, so $U_n \setminus \bigcup_{s \in A_n} C_s \neq \emptyset$. But $U_s \setminus \bigcup_{s \in A_n} C_s$ is open ($\bigcup_{s \in A_n} C_s$ is closed) and there exists $s_1 \in T$ such that $C_{s_1} \subset U_s \setminus \bigcup_{s \in A_n} C_s$. The set $A_n \cup \{s_1\}$ consists of incomparable elements, a contradiction with maximality of A_n.

Let $V_n = \bigcup_{s \in A_n} C_s$. Then $V_n \subset U_n$ and V_n is open. To prove that V_n is dense take arbitrary $t \in T$. A_n is a maximal antichain, so there exists $s \in A_n$ such that $s \le t$ or $t \le s$. In the first case $C_t \subset C_s$, so $C_t \subset V_n$, in the second $C_s \subset C_t$, so $C_s \subset V_n \cap C_t$. In both cases $V_n \cap C_t \neq \emptyset$.

Let $\alpha < \omega_1$ be an ordinal number such that $A_n \subset T_{<\alpha}$ for each $n \in \mathbb{N}$. Let $U = \bigcup_{s \in T_\alpha} C_s$. U is an open dense set, because T_α is a maximal antichain. Moreover, $U \subset \bigcap_{n \in \mathbb{N}} V_n \subset \bigcap_{n \in \mathbb{N}} U_n$. The second inclusion is obvious. To prove the first fix $n \in \mathbb{N}$. If $b \in U$, then obviously there exists $s \in T_\alpha$ such that $b \in C_s$. From the maximality of A_n it follows that there exists $t \in A_n$ comparable with s, and since $A_n \subset T_{<\alpha}$, it must be $t < s$ and $b \in C_s \subset C_t \subset V_n$. □

Corollary 6.7 *If $E \subset X$ is meager, then E is nowhere dense, so it is included in the complement of an open dense set.*

Proof of the corollary follows immediately from the proof of the lemma.

Lemma 6.8 *If $f : X \to \mathbb{R}$ is a function having the Baire property, then there exists $\alpha < \omega_1$ such that for each $s \in T_\alpha$ the function f is constant on C_s.*

Proof of the lemma Let $\mathcal{B} = \{B_n\}_{n \in \mathbb{N}}$ be a countable open basis in $\mathbb{R}$. For each $n \in \mathbb{N}$ the set $f^{-1}(B_n)$ has the Baire property. So there exists an open set W_n such that $W_n \Delta f^{-1}(B_n)$ is meager. Put $U_n = X \setminus \overline{(W_n \Delta f^{-1}(B_n))}$ for $n \in \mathbb{N}$ (here $\overline{X}$ means the closure of X). Each U_n is open and dense in X. Let $A_n \subset T$ be an antichain which is maximal with respect to the property that $C_s \subset W_n$ for each $s \in A_n$. By the proof of Lemma 6.6 there exists an $\alpha < \omega_1$ such that each $A_n \subset T_{<\alpha}$ and if $U = \bigcup\{C_s : s \in T_\alpha\}$ is the open dense set, then U is disjoint from $W_n \triangle f^{-1}(B_n)$ for each $n \in \mathbb{N}$.

We shall prove that f is constant on each C_s for $s \in T_\alpha$. Suppose that this is not the case. Then there exist $b_1, b_2 \in C_s$ such that $f(b_1) \neq f(b_2)$, so there exists $n_0 \in \mathbb{N}$ for which $f(b_1) \in B_{n_0}$ and $f(b_2) \notin B_{n_0}$. Because $b_1 \in f^{-1}(B_{n_0}) \cap U$ and U is disjoint from $W_{n_0} \triangle f^{-1}(B_{n_0})$, we have that $b_1 \in W_{n_0}$. Hence there exists $t \in T_{<\alpha}$ such that $C_t \subset W_{n_0}$ and $b_1 \in C_t$. Since $t < s$, we have also that $b_2 \in C_t$ and $b_2 \in f^{-1}(B_{n_0})$ which contradicts $f(b_2) \notin B_{n_0}$. □

Lemma 6.9 *If $\{f_n\}_{n \in \mathbb{N}}$ is a sequence of real-valued functions defined on X having the Baire property, then there exists an ordinal number $\alpha < \omega_1$ such that for each $s \in T_\alpha$ and each $n \in \mathbb{N}$ the function f_n is constant on C_s.*

Proof of the lemma For each $n \in \mathbb{N}$ the Lemma 6.8 assures the existence of $\alpha_n < \omega$ such that for each $s \in T_{\alpha_n}$ the function f_n is constant on C_s. If α is an ordinal number greater than all α_n's, then it fulfills all requirements. □

Let's return to the proof of theorem. If a set $H \subset X$ has the Baire property and is not meager, so it is of the form $H = U \triangle E$, where U is open and E is meager, then $U \neq \emptyset$. From the proof of Lemma 6.6 it follows that there exists an open set $V \subset H$. If $s \in T$ is such that $C_s \subset V$ and $t_1, t_2 \in T$, $t_1 \neq t_2$, are among immediate successors of s, then $C_{t_1} \cup C_{t_2} \subset C_s$ and $C_{t_1} \cap C_{t_2} = \emptyset$. Hence the pair (Baire sets, meager sets) is not atomic (even atomless).

The convergence $\mathcal{I}$-almost everywhere implies the convergence with respect to $\mathcal{I}$ (in category), so it is sufficient to prove the converse. Let $\{f_n\}_{n \in \mathbb{N}}$ be the sequence of real-valued Baire measurable functions defined on X which converges to a function $f : X \to \mathbb{R}$ with respect to $\mathcal{I}$. By virtue of Lemma 6.9 there exists an ordinal number $\alpha < \omega_1$ such that for each $s \in T_\alpha$ and $n \in \mathbb{N}$ the function f_n is constant on C_s, say $f_n(x) = c_{s,n}$ for $x \in s$.

From the assumptions it follows that for each subsequence $\{c_{s,n_m}\}_{m \in \mathbb{N}}$ there exists a subsequence $\{c_{s,n_{m_p}}\}_{p \in \mathbb{N}}$ which is convergent, so from the Urysohn property it follows that the sequence $\{c_{s,n}\}_{n \in \mathbb{N}}$ converges. Put $c_s = \lim_{n \to \infty} c_{s,n}$. Then the sequence $\{f_n\}_{n \in \mathbb{N}}$ converges on a set $\bigcup_{s \in T_\alpha} C_s$ to a function $f(x) = c_s$ for $x \in C_s$ for each $s \in T_\alpha$, which means the convergence $\mathcal{I}$-a.e., since $\bigcup_{s \in T_\alpha} C_s$ is open and dense. □

Chapter 7
Rearrangements

Bilyen et al. [BKL86] have proved that if a series $\sum_{n=1}^{\infty} f_n$ of real functions of a real variable diverges almost everywhere, then the series $\sum_{n=1}^{\infty} f_{\Pi(n)}$ also diverges a.e. for the overwhelming majority of permutations $\Pi\colon \mathbb{N} \to \mathbb{N}$. In fact, their proof is valid in more general setting. Below we shall present the proof for measure and its category analogue.

Let G be the group of permutations of the set of positive integers. For $\Pi_1, \Pi_2 \in G$ put

$$d(\Pi_1, \Pi_2) = \sum 2^{-n} \left(d_n(\Pi_1, \Pi_2) + d_n(\Pi_1^{-1}, \Pi_2^{-1})\right),$$

where

$$d_n(\Pi_1, \Pi_2) = \frac{|\Pi_1(n) - \Pi_2(n)|}{1 + |\Pi_1(n) - \Pi_2(n)|}.$$

G with this metric is a Polish space.

Let H be the set of all elements of G which are the identity outside of a finite set. H is a countable subgroup of G. Moreover, for each $\Pi_0 \in H$ the function $f\colon G \to G$ defined by $f(\Pi) = \Pi_0 \circ \Pi$ is a homeomorphism. Indeed, let $\Pi_0 \in H$ and let $\{\Pi_n\}_{n\in\mathbb{N}}$ be a sequence of elements of G such that $\lim_{n\to\infty} \Pi_n = \Pi$. Then for each $k \in N$ there exists $n(k)$ such that $\Pi_n(k) = \Pi(k)$ for each $n \geq n(k)$. But then $(\Pi_0 \circ \Pi_n)(k) = (\Pi_0 \circ \Pi)(k)$ for such k and we are done.

Let now $\Pi_0 \in G$. Take the arbitrary open set U from the neighbourhood basis $O(\Pi_1)$. Then there exists $k \in \mathbb{N}$ such that $U \supset \{\Pi \in G\colon \Pi(1) = \Pi_1(1), \ldots \Pi(k) = \Pi_1(k)\}$. It is not difficult to find $\Pi_2 \in H$ such that $\Pi_2 \circ \Pi_0 \in U$.

If X is a Polish space and $\mathcal{F}$—family of homeomorphisms of X, then we say that a set $A \subset X$ is F-invariant if and only if $f(A) = A$ for each $f \in F$ (see [GH55]). The orbit of $x \in X$ is the set $\{f(x)\colon f \in F\}$.

Lemma 7.1 ([BKL86] or [GH55]) *Let X be a Polish space, F—a countable group of homeomorphisms of X with at least one dense orbit and $A \subset X$—an F-invariant subset of X having the Baire property. Then A is either of the first category or residual*

W. Wilczyński, *Convergence in Measure and in Category*, Lecture Notes in Mathematics 2394, https://doi.org/10.1007/978-3-032-23143-7_7

in X. If each orbit is dense and $X \setminus A$ is the union of F-invariant sets of type $\mathcal{G}_\delta$ and A is residual, then $A = X$.

Proof Let $U \subset X$ be an open set such that $A \triangle U$ is of the first category. If $U = \emptyset$ then $A = A \triangle U$ is of the first category. Suppose that $U \neq \emptyset$. If $V = \bigcup_{f \in F} f(U)$, then V is open and dense in X, because F consists of homeomorphisms and has at least one dense orbit. Observe that $A \triangle V$ is of the first category, since

$$A \triangle V \subset \bigcup_{f \in F} (A \triangle f(U)) = \bigcup_{f \in F} (f(A) \triangle f(U)) = \bigcup_{f \in F} (f(A \triangle U))$$

and the last set is of the first category as a countable union of sets of the first category. Now $X \setminus A = A \triangle X \subset (A \triangle V) \cup (X \setminus V)$, so $X \setminus A$ is of the first category as a union of two sets of the first category, which means that A is residual.

To prove the second part suppose that one of the F-invariant $\mathcal{G}_\delta$'s, the summands of $X \setminus A$ is not empty. Then it is dense, since all orbits are dense, so it is residual and $X \setminus A$ is also residual, a contradiction. Hence $X \setminus A = \emptyset$. □

We shall use the terminology and denotations from [BKL86]: a block is a finite sequence of consecutive positive integers. If B is a block and $n \in \mathbb{N}$, then we shall write $n \leq B$ if $j \in B$ implies $n \leq j$.

Lemma 7.2 *Let $K \subset [0, 1]$ be a closed set and $\{f_n\}_{n \in \mathbb{N}}$—a sequence of continuous functions defined on K. Suppose that there exists a number $\delta > 0$ such that for each $x \in K$ and $n_0 \in \mathbb{N}$ there exists a block B such that $n_0 \leq B$ and $|\sum_{j \in B} f_j(x)| > \delta$. Then the set $\{\Pi \in G\colon \sum_{n \in \mathbb{N}} f_{\Pi(n)}(x)$ diverges for each $x \in K\}$ is residual in G.*

Proof If B is a block let $U(B) = \{x \in K : |\sum_{j \in B} f_j(x)| > \delta\}$. The set $U(B)$ is an open subset of K (possibly empty). If $n \in \mathbb{N}$, then there exists a finite family of blocks $\{B_1, B_2, \ldots, B_p\}$ such that $n \leq B_i$, $U(B_i) \neq \emptyset$ for $1 \leq i \leq p$ and $\bigcup_{i=1}^{p} U(B_i) = K$ since K is compact. Observe also that the set

$$\Big\{\Pi \in G : |\sum_{j \in B_i} f_{\Pi(j)}(x)| > \delta \text{ for all } x \in U(B_i)\Big\}$$

is open and nonempty for $1 \leq i \leq p$, because it includes the identity.

Let

$$A = \bigcap_{n \in \mathbb{N}} \bigcup_{\substack{n \leq B_1, \ldots, B_p,\ U(B_i) \neq \emptyset \\ \bigcup_{i=1}^{p} U(B_i) = K}} \bigcap_{i=1}^{p} \Big\{\Pi \in G : \Big|\sum_{j \in B_i} f_{\Pi(j)}(x)\Big| > \delta \text{ for all } x \in U(B_i)\Big\}.$$

A is nonempty F-invariant set of type $\mathcal{G}_\delta$. By virtue of Lemma 7.1 A is residual set in G. This ends the proof, since A is included in the set $\{\Pi \in G\colon \sum_{n \in \mathbb{N}} f_{\Pi(n)}(x)$ diverges for each $x \in K\}$. □

Theorem 7.3 ([BKL86]) *If $\{f_n\}_{n\in\mathbb{N}}$ is a sequence of real-valued measurable functions defined on $[0, 1]$ such that $\sum_{n\in\mathbb{N}} f_n(x)$ diverges almost everywhere, then the set $\{\Pi \in G\colon \sum_{n\in\mathbb{N}} f_{\Pi(n)}(x)$ diverges a.e$\}$ is residual in G.*

Proof Put

$$D_q = \bigcap_{k=1}^{\infty} \bigcup_{\substack{m,n \\ k\le m\le n}} \left\{x \in [0, 1]\colon \left|\sum_{j=m}^{n} f_j(x)\right| > \frac{1}{q}\right\}$$

for $q \in \mathbb{N}$. Each D_q is Lebesgue measurable and

$$\bigcup_{q=1}^{\infty} D_q = \left\{x \in [0, 1]\colon \sum_{n\in\mathbb{N}} f_n(x) \text{ diverges}\right\},$$

so $\lambda\left(\bigcup_{q=1}^{\infty} D_q\right) = \lim_{q\to\infty} \lambda(D_q) = 1$. For each $q \in \mathbb{N}$ the Lusin Theorem implies the existence of a closed set $K_q \subset D_q$ such that $\lambda(D_q \setminus K_q) < \frac{1}{q}$ and each $f_n|_{K_q}$ is continuous. The set $R_q = \left\{\Pi \in G\colon \sum_{n\in\mathbb{N}} f_{\Pi(n)}(x) \text{ diverges for all } x \in K_q\right\}$ is residual in G, so the set $R = \bigcap_{q=1}^{\infty} R_q$ is also residual. Observe that

$$R \subset \{\Pi \in G\colon \sum_{n\in\mathbb{N}} f_{\Pi(n)}(x) \text{ diverges a.e in } [0, 1]\},$$

which ends the proof. □

In [BKL86] one can find also the following category analogue of the last theorem. We shall need the following lemma:

Lemma 7.4 *If $\{x_n\}_{n\in\mathbb{N}}$ is a sequence of real numbers and $A = \left\{\Pi \in G\colon \sum_{n\in\mathbb{N}} x_{\Pi(n)}(x) \text{ converges}\right\}$ then either $A = G$ or A is of the first category.*

Proof Observe that

$$G \setminus A = \bigcup_{k=1}^{\infty} \bigcap_{l=1}^{\infty} \bigcup_{\substack{m,n \\ l\le m\le n}} \left\{\Pi \in G\colon \left|\bigcup_{j=m} x_{\Pi(j)}\right| > \frac{1}{k}\right\}$$

is a countable union of F-invariant sets of type $\mathcal{G}_\delta$. From Lemma 7.1 it follows that A is either of the first category or $A = G$. □

Theorem 7.5 *If $\{f_n\}_{n\in\mathbb{N}}$ is a sequence of a real-valued functions having the Baire property such that the series $\sum_{n=1}^{\infty} f_n(x)$ diverges on a residual subset of $[0, 1]$, then the set $\left\{\Pi \in G\colon \sum_{n\in\mathbb{N}} f_{\Pi(n)}(x) \text{ diverges on a residual subset of } [0, 1]\right\}$ is a residual subset of G.*

Proof For $n \in \mathbb{N}$ consider the function $P_n\colon [0,1] \times G \to \mathbb{R}$ of the form $P_n(x, \Pi) = f_{\Pi(n)}(x)$. Let $U \subset \mathbb{R}$ be an open set. Since $P_n^{-1}(U) = \bigcup_{k=1}^{\infty} \left(f_k^{-1}(U) \times F_k\right)$, where $F_k = \{\Pi \in G\colon \Pi(n) = k\}$, $f^{-1}(U)$ has the Baire property and F_k is closed for each $k \in \mathbb{N}$, the function P_n has the Baire property. Observe that

$$B = \left\{(x,\Pi)\colon \sum_{n\in\mathbb{N}} f_{\Pi(n)}(x) \text{ diverges}\right\} = \bigcup_{k=1}^{\infty}\bigcap_{l=1}^{\infty} \bigcup_{\substack{m,n\\ k\le m\le n}} \left\{(x,\Pi)\colon \left|\sum_{m\le j\le n} f_{\Pi(j)}(x)\right| > \frac{1}{k}\right\},$$

so $B \subset [0,1] \times G$ has the Baire property. From the assumption of theorem and the last lemma it follows that $B_x = \{\Pi\colon (x,\Pi) \in B\}$ is residual in G except for a first category set of x's. Hence B is residual in $[0,1] \times G$ and from Kuratowski-Ulam theorem [Oxt80] it follows that $B^{\Pi} = \{x\colon (x,\Pi) \in B\}$ is a residual subset of $[0,1]$ except a first category set of Π's. □

Remark 7.6 The problem of rearrangements of Fourier series of functions belonging to $L^2(-\pi,\pi)$ was considered by Garsia [Gar64].

Recall the result of Kolmogorov ([Kol24], see also [Marc33]): if $\{s_n\}_{n\in\mathbb{N}}$ are partial sums of the Fourier series of a function $f \in L^2$ and $\frac{n_k}{n_{k+1}} > q > 1$ for $k \in \mathbb{N}$, then $\{s_{n_k}\}_{k\in\mathbb{N}}$ converges to f almost everywhere. Now we know that the situation is much better—Carleson [Car66] has proved that the Fourier series of a function $f \in L^2$ converges to f a.e. solving the classical problem of Lusin. The original theorem of [Gar64] looks as follows: if successively and independently for each $k \in \mathbb{N}$ we permute at random the terms of the Fourier series of $f \in L^2$ whose indices are comprised between n_k and n_{k+1} among themselves, then with probability one the resulting rearranged series will converge almost everywhere. The probability in the subspace of the space of all permutations of $\mathbb{N}$ is defined in the following way: let Ω_k denote the probability space whose elements are the permutations σ_k of the positive integers m which satisfy $n_{k-1} < m \le n_k$ assuming each σ_k equally probable and let $\Omega = \Omega_1 \times \Omega_2 \times \cdots$ denote the Cartesian product of these probability spaces equipped with the product probability measure. To each element $\sigma = (\sigma_1, \sigma_2, \ldots) \in \Omega$ there corresponds a rearrangement of the Fourier series of f. In the same paper it is proved that similar theorem holds for general orthonormal expansions of function from L^2.

Chapter 8
Improving and Spoiling Convergence

Recall three classical theorems concerning series of real numbers.

Theorem A (Abel and Dini [Kno51], Chap. IX, Sect. 39, p. 173) *If* $\sum_{k=1}^{\infty} a_k = \infty$, $a_k \geq 0$ *for* $k \in \mathbb{N}$, *then there exists a sequence* $\{t_k\}_{k\in\mathbb{N}}$ *of positive numbers such that* $\lim_{k\to\infty} t_k = 0$ *and* $\sum_{k=1}^{\infty} t_k a_k = \infty$.

Theorem B (Dini [Kno51], Chap. IX, Sect. 39, p. 175) *If* $\sum_{k=1}^{\infty} a_k < \infty$, $a_k \geq 0$ *for* $k \in \mathbb{N}$, *then there exists a sequence* $\{t_k\}_{k\in\mathbb{N}}$ *of positive numbers such that* $\lim_{k\to\infty} t_k = \infty$ *and* $\sum_{k=1}^{\infty} t_k a_k < \infty$.

Theorem C (Stieltjes [Kno51], Chap. IX, Sect. 41, p. 178) *For each sequence* $\{\delta_k\}_{k\in\mathbb{N}}$ *of positive numbers such that* $\lim_{k\to\infty} \delta_k = 0$ *there exists a series* $\sum_{k=1}^{\infty} a_k$, $a_k > 0$ *for* $k \in \mathbb{N}$, *such that* $\sum_{k=1}^{\infty} a_k = \infty$ *and* $\sum_{k=1}^{\infty} \delta_k a_k < \infty$.

These theorems show that there does not exist neither a series which converges slower than any other series nor a series which diverges slower than any other series. This implies among others that no comparison test can be effective with all series. Theorem A remains true also for a sequence $\{S_i\}_{i\in\mathbb{N}}$ of divergent series $\sum_{k=1}^{\infty} a_{ik}$ of positive numbers. Namely, there exists a sequence $\{t_k\}_{k\in\mathbb{N}}$ independent from i such that all series $\sum_{k=1}^{\infty} t_k a_{ik}$, $i \in \mathbb{N}$, are divergent. It was proved in [Sth28] with the use of Banach-Steinhaus theorem. In the same paper Hugo Steinhaus dealt with the following question of Stanisław Ruziewicz: Suppose that a series of positive functions $\sum_{k=1}^{\infty} a_k(x)$, $a_k : [0, 1] \to \mathbb{R}^+$ for $k \in \mathbb{N}$, is divergent for each $x \in [0, 1]$. Is it possible to find a sequence $\{t_k\}_{k\in\mathbb{N}}$ of positive numbers such that $\lim_{k\to\infty} t_k = 0$ and $\sum_{k=1}^{\infty} t_k a_k(x) = \infty$ for each t? The answer is negative, there exists a sequence $\{a_k(x)\}_{k\in\mathbb{N}}$ of measurable functions such that for each sequence $\{t_k\}_{k\in\mathbb{N}} = T$ there exists a point $x(T) \in [0, 1]$ such that $\sum_{k=1}^{\infty} t_k a_k(x) < \infty$.

W. Wilczyński, *Convergence in Measure and in Category*, Lecture Notes in Mathematics 2394, https://doi.org/10.1007/978-3-032-23143-7_8

Let $\mathbb{T}$ be the set of all positive sequences convergent to 0 ($\text{card}(\mathbb{T}) = \mathfrak{c}$). Take a set $E \subset [0, 1]$, $\lambda(E) = 0$, $\text{card}(E) = \mathfrak{c}$. There exists a function $t \colon \mathbb{T} \xrightarrow{1-1} E$. For each $T = \{t_k\}_{k\in\mathbb{N}}$ there exists $\{a_k{}^T\}_{k\in\mathbb{N}}$ such that $\sum_{k=1}^{\infty} a_k{}^T = \infty$, $a_k{}^T > 0$, and $\sum_{k=1}^{\infty} t_k a_k{}^T < \infty$ (again Stieltjes). Put $a_k(x) = a_k{}^T$ for $t = t(T)$ and $a_k(x) = \frac{1}{k}$ for $t \notin E$, $k \in \mathbb{N}$. All functions a_k are measurable ($= \frac{1}{k}$ a.e.) and

$$\sum_{k=1}^{\infty} a_k(x) = \infty \text{ for each } t \in [0, 1],$$

$$\sum_{k=1}^{\infty} t_k a_k(x) < \infty \text{ for } t = t(T),\ \ T = \{t_k\}_{k\in\mathbb{N}}.$$

If one requires only divergence or convergence almost everywhere, the corresponding versions of Theorems A, B, C remain true [Sth28].

Theorem 8.1 ([Sth28]) *If $\{f_k\}_{k\in\mathbb{N}}$ is a sequence of measurable non-negative functions defined on $[0, 1]$ such that $\sum_{k=1}^{\infty} f_k = +\infty$ almost everywhere, then there exists a sequence $\{t_k\}_{k\in\mathbb{N}}$ of positive numbers such that $\lim_{k\to\infty} t_k = 0$ and $\sum_{k=1}^{\infty} t_k f_k(x) = +\infty$ almost everywhere on $[0, 1]$.*

Proof Let $\{\varepsilon_k\}_{k\in\mathbb{N}}$ be a sequence of positive numbers such that $\sum_{k=1}^{\infty} \varepsilon_k < +\infty$. There exists an increasing sequence $\{n_i\}_{i\in\mathbb{N}}$ of positive integers such that for $i \in \mathbb{N}$ we have $\lambda(E_i) = \lambda\left([0, 1] \setminus \left\{x \in [0, 1] \colon \sum_{n_{i-1}+1}^{n_i} f_k(x) \geq i\right\}\right) < \varepsilon_i$ (here $n_0 = 0$). Put $t_k = 1$ for $1 \leq k \leq n_1$ and $t_k = \frac{1}{i}$ for $n_{i-1} \leq k \leq n_i$. Obviously $\lim_{k\to\infty} t_k = 0$ and $\sum_{n_i+1}^{n_{i+j}} t_k f_k(x) \geq j$ except on a set $\bigcup_{k=1}^{j} E_{i+k}$. If $F_i = \bigcup_{k=i+1}^{\infty} E_k$, then $\sum_{k=n_i+1}^{\infty} t_k f_k(x) = \infty$ for $x \notin F_i$, so also $\sum_{k=i+1}^{\infty} t_k f_k(x) = \infty$. But $\lambda(F_i) \leq \sum_{k=i+1}^{\infty} \varepsilon_k$, and $\lim_{i\to\infty} \sum_{k=i+1}^{\infty} \varepsilon_k = 0$, which proves the theorem.

Theorem 8.2 *If $\{f_k\}_{k\in\mathbb{N}}$ is a sequence of measurable non-negative functions defined on $[0, 1]$ and $\sum_{k=1}^{\infty} f_k(x) < +\infty$ almost everywhere, then there exists a sequence $\{t_k\}_{k\in\mathbb{N}}$ of positive numbers such that $\lim_{k\to\infty} t_k = \infty$ and $\sum_{k=1}^{\infty} t_k f_k(x) < +\infty$ almost everywhere.*

Proof Let again $\{\varepsilon_k\}_{k\in\mathbb{N}}$ be a sequence of positive numbers such that $\sum_{k=1}^{\infty} \varepsilon < +\infty$. There exists an increasing sequence $\{n_i\}_{i\in\mathbb{N}}$ of positive integers such that if $F_i = \left\{x \in [0, 1] \colon \left|\sum_{k=m}^{p} f_k(x)\right| < \frac{i}{i^3} \text{for each } m, p > n_i\right\}$ and $E_i = [0, 1] \setminus F_i$, then $\lambda(E_i) < \varepsilon_i$. Put $t_k = i$ for $n_i < k \leq n_{i+1}$, $i \in \mathbb{N}$. If $m > n_i$, $p > m$ and $n_{i+j} < p \leq n_{i+j+1}$, then

$$\left|\sum_{k=m}^{p} t_k f_k(x)\right| \le \left|\sum_{k=m}^{n_{i+1}} t_k f_k(x)\right| + \left|\sum_{k=n_{i+1}}^{n_{i+2}} t_k f_k(x)\right| + \dots$$

$$+ \left|\sum_{k=n_{i+j+1}}^{n_{i+j}} t_k f_k(x)\right| + \left|\sum_{k=n_{i+j}+1}^{p} t_k f_k(x)\right| \le$$

$$\le \frac{1}{i^2} + \frac{1}{(i+1)^2} + \dots + \frac{1}{(i+j)^2} < \frac{1}{i-1}$$

for $x \notin \bigcup_{k=i}^{\infty} E_k$. From this it follows that the series $\sum_{k=1}^{\infty} t_k f_k(x)$ converges everywhere except on a set of arbitrary small measure, hence almost everywhere.

Theorem 8.3 *If $\{f_k\}_{k\in\mathbb{N}}$ is a sequence of measurable functions defined on $[0, 1]$ such that $\lim_{k\to\infty} f_k = 0$ almost everywhere, then there exists a sequence $\{t_k\}_{k\in\mathbb{N}}$ of positive numbers such that $\lim_{k\to\infty} t_k = \infty$ and $\sum_{k=1}^{\infty} t_k f_k(x)$ converges almost everywhere.*

Proof Let $\{\varepsilon_k\}_{k\in\mathbb{N}}$ be a sequence of positive numbers such that $\sum_{k=1}^{\infty} \varepsilon < +\infty$. There exists an increasing sequence $\{n_i\}_{i\in\mathbb{N}}$ of positive integers such that if $F_i = \{x \in [0, 1]\colon\ |f_k(x)| < \varepsilon_i \text{ for } n > n_i\}$ and $E_i = [0, 1] \setminus F_i$, then $\lambda(E_i) < \varepsilon_i$. Put $d_k = \varepsilon_i$ for $n_i < k \le n_{i+1}$ and $d_k = 1$ for $k \le n_1$. Obviously $\lim_{k\to\infty} d_k = 0$. From Theorem C it follows that there exists a sequence $\{t_k\}_{k\in\mathbb{N}}$ of positive numbers such that $\lim_{k\to\infty} t_k = \infty$ and $\sum_{k=1}^{\infty} t_k d_k < \infty$. It is clear that $\lambda\left(\limsup_{i\to\infty} E_i\right) = 0$. If $x \notin \limsup_{i\to\infty} E_i$, then there exists $i_0 \in \mathbb{N}$ such that for $i \ge i_0$, $x \in F_i$. Hence $\sum_{k=n_{i_0}}^{\infty} |t_k f_k(x)| < \sum_{k=n_{i_0}}^{\infty} t_k d_k$, so $\sum_{k=1}^{\infty} t_k f_k$ converges absolutely almost everywhere. □

We shall prove that Theorem 8.1 has satisfactory analogue for the Baire property while Theorems 8.2 and 8.3 do not.

Theorem 8.4 ([Wil19]) *If $\{f_k(x)\}_{k\in\mathbb{N}}$ is a sequence of functions with the Baire property, $f_k\colon [0, 1] \to \mathbb{R}_+$ for $k \in \mathbb{N}$ and $\sum_{k=1}^{\infty} f_k(x) = \infty$ everywhere except a set of the first category on $[0, 1]$, then there exists a sequence $\{t_k\}_{k\in\mathbb{N}}$ of positive numbers such that $\lim_{k\to\infty} t_k = 0$ and $\sum_{k=1}^{\infty} t_k f_k(x) = \infty$ everywhere except a set of the first category.*

Proof We shall need a simple lemma.

Lemma 8.5 *If $\{f_n\}_{n\in\mathbb{N}}$ is an increasing sequence of functions with the Baire property, $f_n\colon [0, 1] \to \mathbb{R}_+$ for $n \in \mathbb{N}$ and $\lim_{n\to\infty} f_n(x) = \infty$ everywhere except a set of the first category, then for each $p > 0$ and for each interval $[a, b] \subset [0, 1]$ there exists $n_0 \in \mathbb{N}$ such that $E_n = \{x \in [a, b]\colon f_n(x) \ge p\}$ for $n > n_0$ is of the second category (so E_n is residual in some subinterval $[c, d] \subset [a, b]$ as a set having the Baire property).*

Proof of the lemma Let $E_n = \{x \in [a, b]\colon f_n(x) \geq p\}$ for $n \in \mathbb{N}$. Since $\bigcup_{n=1}^{\infty} E_n$ is residual in $[a, b]$, there exists $n_0 \in \mathbb{N}$ such that E_{n_0} is of the second category, each E_n for $n > n_0$ is also of the second category because the sequence $\{E_n\}_{n\in\mathbb{N}}$ is ascending. □

Now, the existence of an increasing sequence of positive integers $\{n_i\}_{i\in\mathbb{N}}$ will follow from the lemma. Let n_1 be a positive integer for which the set $A_1 = \left\{x \in [0, 1]\colon \sum_{k=1}^{\infty} f_k(x) \geq 1\right\}$ is of second category. Let $n_2 > n_1$ be a positive integer for which both sets $A_2^1 = \left\{x \in [0, \frac{1}{2}]\colon \sum_{k=n_1+1}^{n_2} f_k(x) \geq 2\right\}$ and $A_2^2 = \left\{x \in [\frac{1}{2}, 1]\colon \sum_{k=n_1+1}^{n_2} f_k(x) \geq 2\right\}$ are of the second category. Suppose that we have chosen $n_1 < n_2 < \ldots < n_i$. Let $n_{i+1} > n_i$ be a positive integer for which all sets

$$A_{i+1}^j = \left\{x \in [\frac{j-1}{2^i}, \frac{j}{2^i}]\colon \sum_{k=n_i+1}^{n_{i+1}} f_k(x) \geq i+1\right\},$$

$j \in \{1, 2, \ldots, 2^i\}$, are of second category.

Now put $t_k = 1$ for $1 \leq k \leq n_1$ and $t_k = \frac{1}{k+1}$ for $n_i + 1 \leq k \leq n_{i+1}$. We obviously have $\lim_{k\to\infty} t_k == 0$. At the same time $\sum_{k=n_i+1}^{n_{i+1}} t_k f_k(x) \geq 1$ for $\bigcup_{j=1}^{2^i} A_{i+1}^j$. Hence $\sum_{k=1}^{\infty} t_k f_k(x) = \infty$ for $x \in \limsup_{i\to\infty} \left(\bigcup_{j=1}^{2^i} A_{i+1}^j\right) = A$. From the construction it follows that A is residual in $[0, 1]$. □

To show that analogues of Theorems 8.2 and 8.3 do not hold for the Baire property we shall construct a sequence $\{f_n\}_{n\in\mathbb{N}}$ of functions with the Baire property which converges pointwise to zero and does not converge uniformly on any set of the second category with the Baire property. One can find an example of such sequence of functions in [Oxt80], Chap. 8, consisting of continuous functions, but in our construction the sequence is nonincreasing, which enables us to build a series of functions with non-negative terms. Let $I_i^n = \left(\frac{i-1}{2^n}, \frac{i}{2^n}\right)$ for $n \in \mathbb{N}$ and $i \in \{1, 2, \ldots, 2^n\}$, $A_n = \bigcup_{i=1}^{2^{n-1}} A_{2i-1}^n$, $B_n = \bigcup_{i=1}^{2^{n-1}} A_{2i}^n$ for $n \in \mathbb{N}$. Put

$$f_1(x) = \begin{cases} 1 & \text{for } x \in A_1, \\ \frac{1}{2} & \text{for } x \in B_1, \\ 0 & \text{for the remaining } x \in [0, 1]. \end{cases}$$

If $f_1, f_2, \ldots, f_{n-1}$ are already defined, put

$$f_n(x) = \begin{cases} f_{n-1} & \text{for } x \in A_n, \\ \frac{1}{n+1} & \text{for } x \in B_n, \\ 0 & \text{for the remaining } x \in [0, 1]. \end{cases}$$

Obviously $f_{n+1} \leq f_n(x)$ for each $n \in \mathbb{N}$ and $x \in [0, 1]$. We shall show that $\lim_{n\to\infty} f_n(x) = 0$ for each $x \in [0, 1]$. It is obvious if the binary expansion of x contains a finite number of 1's (or equivalently, a finite number of 0's), since then

$f_n(x) = 0$ for sufficiently big $n \in \mathbb{N}$. Suppose now that the binary expansion of x contains infinitely many 0's as well as infinitely many 1's. If $x = (0, a_1 a_2 \dots a_k \dots)_2$ and $a_k = 1$, then from the definition of f_k it follows that $f_k(x) = \frac{1}{k+1}$; if $f_k = 0$, then $f_k(x) = \frac{1}{j+1}$, where $j = \max\{i : a_i = 1 \text{ and } i < k\}$ and $j \to \infty$ as $k \to \infty$. Hence $\lim_{n\to\infty} f_n(x) = 0$ for all $x \in [0, 1]$.

Suppose now that $E \subset [0, 1]$ is the set of the second category having the Baire property. Then there exists an interval $(a, b) \subset [0, 1]$ and a set P of the first category such that $(a, b) \setminus P \subset E$. We shall show that $\{f_n\}_{n\in\mathbb{N}}$ is not uniformly convergent on $(a, b) \setminus P$. Indeed, there exists $n_0 \in \mathbb{N}$ such that at least one component of B_{n_0} is included in $[a, b]$, so there exists a point $x_{n_0} \in B_{n_0} \cap ((a, b) \setminus P))$ and $f_{n_0}(x_{n_0}) = \frac{1}{n_0+1}$. From the construction of $\{f_n\}_{n\in\mathbb{N}}$ it follows that for each $n > n_0$ at least one component of A_n is included in (a, b), so for each $n > n_0$ there exists a point $x_n \in A_n \cap ((a, b) \setminus P)$ such that $f_n(x_n) = \frac{1}{n_0+1}$. Hence $\{f_n\}_{n\in\mathbb{N}}$ is not uniformly convergent on $(a, b) \setminus P$ and obviously on E. □

Theorem 8.6 ([Wil19]) *There exists a sequence $\{g_k\}_{k\in\mathbb{N}}$ of functions having the Baire property, $g_k : [0, 1] \to \mathbb{R}^+$ for $k \in \mathbb{N}$ and $\sum_{k=1}^{\infty} g_k(x) < \infty$ for each $x \in [0, 1]$ such that for each sequence $\{t_k\}_{k\in\mathbb{N}}$ of positive numbers tending to infinity the series $\sum_{k=1}^{\infty} t_k g_k(x)$ is divergent to infinity on a set residual in* $[0, 1]$.

Proof Put $g_1(x) = 1 - f_1(x)$ and $g_k(x) = f_{k-1}(x) - f_k(x)$ for $k \geq 2$, where $\{f_n\}_{n\in\mathbb{N}}$ is the sequence constructed earlier. Then $\sum_{i=1}^{k} g_i(x) = 1 - f_k(x)$ for $k \in \mathbb{N}$, so $\sum_{k=1}^{\infty} g_k(x) = 1$ except on a denumerable set. Observe that $g_k(x) \geq \frac{1}{k} - \frac{1}{k+1} = \frac{1}{k(k+1)}$ for $t \in B_k$, since $f_{k-1}(x) \geq \frac{1}{k}$ and $f_k(x) = \frac{1}{k+1}$ on B_k. From the construction of $\{f_n\}_{n\in\mathbb{N}}$ it follows immediately that $a_{k+p}(x) \geq \frac{1}{k(k+1)}$ on $A_{k+p} \cap B_k$ for each $p \in \mathbb{N}$.

Let $\{p_k\}_{k\in\mathbb{N}}$ be an increasing sequence of positive integers such that $t_{k+p_k} > k(k+1)$ for each $k \in \mathbb{N}$. Then we have $t_{k+p_k} \cdot g_{k+p_k}(x) > 1$ for $t \in A_{k+p_k} \cap B_k$. Hence $\sum_{k=1}^{\infty} t_k g_k(x) \geq \sum_{k=1}^{\infty} t_{k+p_k} g_{k+p_k}(x) = +\infty$ for $x \in \limsup_{k\to\infty}(A_{k+p_k} \cap B_k)$ and this last set is residual in $[0, 1]$. □

Theorem 8.7 ([Wil19]) *There exists a sequence $\{f_k(x)\}_{k\in\mathbb{N}}$ of functions having the Baire property, $f_k : [0, 1] \to \mathbb{R}^+$ and $\lim_{k\to\infty} f_k(x) = 0$ for $x \in [0, 1]$ such that for each sequence $\{a_k\}_{k\in\mathbb{N}}$ of positive numbers for which $\sum_{k=1}^{\infty} a_k = +\infty$ the series $\sum_{k=1}^{\infty} a_k f_k(x)$ is divergent to infinity on a set residual in* $[0, 1]$.

Proof Let $\{f_k\}_{k\in\mathbb{N}}$ be a sequence of functions constructed earlier. Suppose that $\sum_{k=1}^{\infty} a_k = +\infty$. There exists $n_1 \in \mathbb{N}$ such that $\sum_{k=1}^{n_1} a_k > 1$. Observe that $f_1(x) = f_2(x) = \cdots = f_{n_1}(x) = 1$ for $t \in (0, \frac{1}{2^{n_1}})$, so $\sum_{k=1}^{n_1} a_k f_k(x) > 1$ on this interval. There exists $n_2 > n_1$ such that $\sum_{k=n_1+1}^{n_2} a_k > 2$. Again observe that $f_{n_1+1}(x) \geq \frac{1}{2}, f_{n_1+2}(x) \geq \frac{1}{2}, \dots, f_{n_2}(x) \geq \frac{1}{2}$ for $x \in (0, \frac{1}{2^{n_2}}) \cup (\frac{1}{2}, \frac{1}{2} + \frac{1}{2^{n_2}})$ (in fact the value of all functions above is equal to 1 in the first interval and to $\frac{1}{2}$ in the second), so $\sum_{k=n_1+1}^{n_2} a_k f_k(x) > 1$ on this set. Suppose that we have found $n_1 < n_2 < \cdots < n_i$. Let n_{i+1} be such number that $\sum_{k=n_i+1}^{n_{i+1}} a_k > i + 1$. Similarly as before we observe that $f_{n_i+1}(x) \geq \frac{1}{i+1}$, $f_{n_i+2}(x) \geq \frac{1}{i+1}, \dots, f_{n_{i+1}}(x) \geq \frac{1}{i+1}$ on the open set E_i such that the intersection of E_i with each component of A_{i+1} is nonempty (namely it is equal to the interval $(\frac{2(j-1)}{2^{i+1}}, \frac{2j-1}{2^{i+1}}) \cap (\frac{2(j-1)}{2^{i+1}}, \frac{2(j-1)}{2^{i+1}} + \frac{1}{2^{n_{i+1}}})$). Hence $\sum_{k=n_i+1}^{n_{i+1}} a_k f_k(x) > 1$ for $t \in E_i$. Finally $\sum_{k=1}^{\infty} a_k f_k(x) = +\infty$ for $x \in \limsup_{i\to\infty} E_i$ and this last set is residual on $[0, 1]$. □

Definition 8.8 We say that the sequence $\{f_n\}_{n\in\mathbb{N}}$ of real-valued measurable functions defined on $[0, 1]$ converges almost uniformly to a function f if and only if there exists a sequence $\{B_i\}_{i\in\mathbb{N}}$ of measurable sets such that $m(\bigcup_{i=1}^{\infty} B_i) = 1$ ($[0, 1] \setminus \bigcup_{i=1}^{\infty} B_i$ is a nullset) and $\{f_n|_{B_i}\}_{n\in\mathbb{N}}$ converges uniformly to $\{f|_{B_i}\}$ ($f_n|_{B_i} \rightrightarrows f|_{B_i}$) for each $i \in \mathbb{N}$ (converges to f in the sense of Egorov, see Theorem 1.27).

Definition 8.9 We say that the sequence $\{f_n\}_{n\in\mathbb{N}}$ of measurable functions satisfies the vanishing restriction with respect to f if for all $\alpha > 0$ we have

$$\lim_{n\to\infty} m(E_n^f(\alpha)) = 0, \quad \text{where } E_n^f(\alpha) = \bigcup_{j=n}^{\infty} \{x \in [0, 1] : |f_j(x) - f(x)| > \alpha\}.$$

Since the sequence $\{E_n^f(\alpha)\}_{n\in\mathbb{N}}$ is decreasing, the last condition means that $\bigcap_{n=1}^{\infty} E_n^f(\alpha)$ is a nullset.

Theorem 8.10 ([Bar80]) *If the sequence $\{f_n\}_{n\in\mathbb{N}}$ of measurable functions defined on $[0, 1]$ converges almost uniformly to f, then it satisfies the vanishing restriction with respect to f. If $\{f_n\}_{n\in\mathbb{N}}$ satisfies the vanishing restriction with respect to f, then $\{f_n\}_{n\in\mathbb{N}}$ converges a.e. to f, so also almost uniformly to f.*

Proof Suppose that $\{f_n\}_{n\in\mathbb{N}}$ converges almost uniformly fo f and $\{B_i\}_{i\in\mathbb{N}}$ is a sequence of sets described in Definition 8.8. Fix $\alpha > 0$ and $\varepsilon > 0$. Then for each $i \in \mathbb{N}$ there exists $n_i \in \mathbb{N}$ such that $\{x \in [0, 1]\colon |f_n(x) - f(x)| > \alpha\} \cap B_i = \emptyset$ for each $n \geq n_i$. There exists $i_0 \in \mathbb{N}$ such that $\lambda\left(\bigcup_{i=1}^{i_0} B_i\right) > 1 - \varepsilon$. If we put $n_0 = \max(n_1, n_2, \dots, n_{i_0})$, then $\{x \in [0, 1]\colon |f_n(x) - f(x)| > \alpha\} \cap \bigcup_{i=1}^{i_0} B_i = \emptyset$ for $n \geq n_0$. Hence $E_{n_0}^f(\alpha) \cap \bigcup_{i=1}^{i_0} B_i = \emptyset$ and $\lambda\left(E_{n_0}^+(\alpha)\right) < \varepsilon$ for $n \geq n_0$, so $\lim_{n\to\infty} \lambda\left(E^f(\alpha)\right) = 0$.

Suppose now, that $\{f_n\}_{n\in\mathbb{N}}$ does not converge a.e. to f. Then

$$\lambda\left(\{x \in [0, 1]\colon \limsup_{n\to\infty} |f_n(x) - f(x)| > 0\}\right) > 0.$$

There exists $\alpha > 0$ for which $\lambda\left(\{x \in [0, 1]\colon \limsup_{n\to\infty} |f_n(x) - f(x)| > \alpha\}\right) > 0$. Observe that if $\limsup_{n\to\infty} |f_n(x_0) - f(x_0)| > \alpha$, then

$$x_0 \in \limsup_{n\to\infty} \left\{x \in [0, 1]\colon |f_n(x) - f(x)| > \frac{\alpha}{2}\right\} = \bigcap_{n=1}^{\infty} E_{n_0}^f\left(\frac{\alpha}{2}\right).$$

Hence $\lambda\left(E^f_{n_0}(\frac{\alpha}{2})\right) \geq \lambda\left(\{x \in [0,1]\colon \limsup_{n\to\infty} |f_n(x) - f(x)| > \alpha\}\right) > 0$ and $\{f_n\}_{n\in\mathbb{N}}$ does not satisfy the vanishing restriction with respect to f. □

Definition 8.11 We shall say that the sequence $\{f_n\}_{n\in\mathbb{N}}$ of functions having the Baire property converges to f B-almost uniformly if there exist a sequence $\{B_i\}_{i\in\mathbb{N}}$ of sets having the Baire property such that $[0,1] \setminus \bigcup_{i=1}^{\infty} B_i$ is of the first category and $f_{n|B_i} \rightrightarrows f_{|B_i}$ for each $i \in \mathbb{N}$. We can (and shall) suppose that the sequence $\{B_i\}_{i\in\mathbb{N}}$ is ascending.

Definition 8.12 We shall say that the sequence $\{f_n\}_{n\in\mathbb{N}}$ satisfies the B-vanishing restriction with respect to f if for all $\alpha > 0$ the set $\bigcap_{n=1}^{\infty} E^f_n(\alpha)$ is of the first category, where $E^f_n(\alpha)$ is the same as above.

Theorem 8.13 *If the sequence $\{f_n\}_{n\in\mathbb{N}}$ of functions having the Baire property defined on* $[0,1]$ *converges B-almost uniformly to f, then if satisfies the B-vanishing condition with respect to f. There exists a sequence $\{f_n\}_{n\in\mathbb{N}}$ of functions having the Baire property defined on* $[0,1]$ *which converges everywhere to $f \equiv 0$ and satisfies the B-vanishing restriction with respect to f but which does not converge B-almost uniformly to f.*

Proof Suppose that $\{B_i\}_{i\in\mathbb{N}}$ is an increasing sequence of sets having the Baire property such that $[0,1] \setminus \bigcup_{i=1}^{\infty} B_i$ is of the first category and $f_{n|B_i} \rightrightarrows f_{|B_i}$ for each $i \in \mathbb{N}$.

Fix $\alpha > 0$. For each $i \in \mathbb{N}$ there exists $n_{\alpha,i} \in \mathbb{N}$ such that for each $x \in B_i$ and for each $n \geq n_{\alpha,i}$ we have $|f_n(x) - f(x)| \leq \alpha$. So $E_{n_{\alpha,i}}(\alpha) \subset [0,1] \setminus B_i$ and $\bigcap_{n=1}^{\infty} E_n(\alpha) \subset \bigcap_{i=1}^{\infty}([0,1] \setminus B_i) = [0,1] \setminus \bigcup_{i=1}^{\infty} B_i$ which ends the proof of the first part of the theorem.

To prove the second part it is sufficient to observe that the sequence $\{f_n\}_{n\in\mathbb{N}}$ of functions constructed before the proof of Theorem 8.6 satisfies the B-vanishing restriction. Simultaneously, if this sequence converges uniformly to the zero function on a set $E \subset [0,1]$ having the Baire property, then E is of the first category. Hence it is clear that $\{f_n\}_{n\in\mathbb{N}}$ does not converge B-almost uniformly to the zero function. □

Chapter 9
Translations of Measurable Sets

It is well known that if $\{x_n\}_{n\in\mathbb{N}}$ is a sequence of real numbers convergent to zero and $A \subset [0, 1]$ is a measurable set then $\lim_{n\to\infty} \lambda(A \cap (A + x_n)) = \lambda(A)$ (and, of course, $\lim_{n\to\infty} \lambda(A\triangle(A + x_n)) = 0$). In [Bari61] on p. 901 one can find a little bit stronger result: $\lim_{n\to\infty} \lambda\big((A - x_n) \cap A \cap (A + x_n)\big) = \lambda(A)$ proved with the use of integral modulus of continuity. We shall prove that $\lim_{n\to\infty} \lambda(A\triangle(A + x_n)) = 0$ using the fact that the function $\rho(A, B) = \lambda(A\triangle B)$ defined for measurable sets A, B is pseudometric (see for example [HS69], Ex. 10.45, p. 144).

Theorem 9.1 *If $A \subset [0, 1]$ is a measurable set and $\{x_n\}_{n\in\mathbb{N}}$ is a sequence convergent to zero, then $\lim_{n\to\infty} \lambda(A\triangle(A + x_n)) = 0$*

Proof Suppose that $\lambda(A) > 0$. If $\lambda(A) = 0$ there is nothing to prove. Take $\varepsilon > 0$. There exists a perfect set $P \subset A$ such that $\rho(A, P) = \lambda(A\triangle P) < \frac{\varepsilon}{6}$. Let $G \supset P$ be an open set such that $\rho(P, G) = \lambda(P\triangle G) < \frac{\varepsilon}{6}$. We can assume that the number of components of G is finite, say k, since P is compact. Then $\rho(A, G) \leq \rho(A, P) + \rho(P, G) < \frac{\varepsilon}{6} + \frac{\varepsilon}{6} = \frac{\varepsilon}{3}$. Obviously also $\rho(A + x_n, G + x_n) < \frac{\varepsilon}{3}$ for each $n \in \mathbb{N}$. Let $n_0 \in \mathbb{N}$ be such that $x_n < \frac{\varepsilon}{6k}$ for $n \geq n_0$. Then $\rho(G, G + x_n) \leq 2k \cdot |x_n| < \frac{\varepsilon}{3}$. Hence

$$\lambda(A\triangle(A + x_n)) = \rho(A, A + x_n) \leq \rho(A, G) + \rho(G, G + x_n) + \rho(G + x_n, A + x_n) <$$
$$< \frac{\varepsilon}{3} + \frac{\varepsilon}{3} + \frac{\varepsilon}{3} = \varepsilon$$

for $n \geq n_0$. □

Corresponding result for sets having the Baire property looks as follows:

Theorem 9.2 *If $A \subset [0, 1]$ is a set having the Baire property, $\{x_n\}_{n\in\mathbb{N}}$ is a sequence convergent to zero, $f = \chi_A$, $f_n = \chi_{A+x_n}$ for $n \in \mathbb{N}$, then $\lim_{n\to\infty} f_n(x) = f(x)$ $\mathcal{K}$-almost everywhere.*

Proof The set A is of the form $A = (G \cup P_1) \setminus P_2$, where G is open, P_1 and P_2 are of the first category. Put

W. Wilczyński, *Convergence in Measure and in Category*, Lecture Notes in Mathematics 2394, https://doi.org/10.1007/978-3-032-23143-7_9

$$P = \mathrm{Fr}\, G \cup P_1 \cup P_2 \cup \bigcup_{n=1}^{\infty}(P_1 + x_n) \cup \bigcup_{n=1}^{\infty}(P_2 + x_n).$$

Obviously P as a countable union of sets of the first category is also of the first category. We shall show that $\lim_{n\to\infty} f_n(x) = f(x)$ for each $x \in [0, 1] \setminus P$. Suppose that $x \in \big([0, 1] \setminus P\big) \cap A$. Then $f(x) = 1$ and $x \in G$. There exists $n_0 \in \mathbb{N}$ such that $x \in G + x_n$ for $n \geq n_0$. Also for these n we have $x \notin P_2 + x_n$. Hence $x \in \big((G + x_n) \cup (P_1 + x_n)\big) \setminus (P_2 + x_n) = A + x_n$ and $f_n(x) = 1$ for $n \geq n_0$. Suppose now that $x \in \big([0, 1] \setminus P\big) \setminus A$. Then $f(x) = 0$ and $x \notin \overline{G}$, so there exists $n_1 \in \mathbb{N}$ such that $x \notin G + x_n$ for $n \geq n_1$. Also for these n we have $x \notin P_1 + x_n$. Hence $x \notin \big((G + x_n) \cup (P_1 + x_n)\big) \setminus (P_2 + x_n) = A + x_n$ and $f_n(x) = 0$ for $n \geq n_1$. □

Observe that if $f = \chi_A$, $f_n = \chi_{A+x_n}$ for $n \in \mathbb{N}$, then in the measurable case we have only convergence in measure while in the second case the convergence is except the set of the first category. The natural question is whether the convergence in measure can be replaced by the convergence almost everywhere. Following [WK92] we shall show that in general it is impossible.

Theorem 9.3 ([WK92]) *There exist a measurable set $A \subset [0, 1]$ and a sequence $\{x_n\}_{n\in\mathbb{N}}$ convergent to zero such that the sequence $\{f_n\}_{n\in\mathbb{N}}$ defined above does not converge to $f = \chi_A$ a.e.*

Proof We shall need the following lemma. □

Lemma 9.4 *If $A \subset [0, 1]$ is a measurable set such that $\lambda((a, b) \cap A) > 0$ and $\lambda((a, b) \setminus A) > 0$ for each interval $(a, b) \subset [0, 1]$, then there exists a sequence $\{x_n\}_{n\in\mathbb{N}}$ convergent to zero such that $\lambda\big(\limsup_{n\to\infty}(A + x_n) \cap [0, 1]\big) = 1$*

Proof of the lemma. First we shall show that for each positive integer m there exists a finite sequence $\left\{h_0^{(m)}, h_1^{(m)}, \ldots h_{k_m}^{(m)}\right\}$ such that $|h_j^{(m)}| < \frac{1}{m}$ for $j \in \{0, 1, \ldots k_m\}$ and $\lambda\left(\bigcup_{j=0}^{k_m}\left(A + h_j^{(m)}\right)\right) > (1 - \frac{1}{m})^2$. From the assumption it follows that the set of points of density of A is dense in $[0, 1]$. Let $0 = x_0 < x_1 < x_2 < \ldots < x_{p_m} < x_{p_m+1} = 1$ be a sequence such that $x_i - x_{i-1} < \frac{1}{m}$ for $i \in \{1, \ldots p_m + 1\}$ and $x_1, x_2, \ldots, x_{p_m}$ are points of density of A. Let $d > 0$ be sufficiently small number for which

$$\frac{\lambda(A \cap [x_i - d, x_i + d])}{2d} > 1 - \frac{1}{m}$$

for $i \in \{1, \ldots p_m\}$. Taking if necessary smaller d we shall assume that $d < \frac{1}{4}\min\left\{x_i - x_{i-1} : i \in \{1, \ldots p_m + 1\}\right\}$ and $2d \cdot (p_m + 1) < \frac{1}{m}$.

Let

$$q_i = \left[\frac{(x_{i-1} - d) - (x_i + d)}{2d}\right] \text{ for } i \in \{1, \dots p_m - 1\},$$

$$q_{p_m} = \left[\frac{1 - (x_{p_m} + d)}{2d}\right] \text{ and } q_0 = \left[\frac{x_1 - d}{2d}\right],$$

where [ ] denotes the integer part. We have $q_i \geq 1$ for $i \in \{1, \dots p_m\}$. Put $q = \max\{q_i : i \in \{1, \dots p_m\}\}$ and let $h_0^{(m)} = 0$, $h_1^{(m)} = 2d$, $h_2^{(m)} = 2 \cdot 2d = 4d, \dots$, $h_q^{(m)} = q \cdot 2d$, $h_{q+1}^{(m)} = -2d$, $h_{q+2}^{(m)} = -2 \cdot 2d = -4d$, ..., $h_{q+q_0}^{(m)} = -q_0 \cdot 2d$. (It is clear that q and q_0 depend on m as well as on the choice of points $x_1, \dots, x_{p_m}$, but we omit this to simplify the denotation). Observe that if $B = \bigcup_{i=1}^{p_m} [x_i - d, x_i + d]$ and $E = \bigcup_{j=0}^{q+q_0} (B + h_j^{(m)})$, then E is a union of at most p_m disjoint closed intervals such that $[0, 1] \setminus E$ consists of at most $(p_m + 1)$ open (among them at most two are half-open) intervals each of the length less than $2d$. Hence $\lambda(E) > 1 - 2d \cdot (p_m + 1) > 1 - \frac{1}{m}$. On each interval $[x_i - d, x_i + d] + h_j^{(m)}$ for $i \in \{1, \dots p_m\}$ and $j \in \{0, 1, \dots q + q_0\}$ we have

$$\lambda\big((A + h_j^{(m)}) \cap \big([x_i - d, x_i + d] + h_j^{(m)}\big)\big) > (1 - \frac{1}{m}) \cdot 2d,$$

so finally

$$\lambda\left(\bigcup_{j=0}^{q+q_0} (A + h_j^{(m)})\right) > \left(1 - \frac{1}{m}\right)^2.$$

Let $\{x_n\}_{n\in\mathbb{N}}$ be a sequence consisting of all numbers $h_j^{(m)}$ for $m \in \mathbb{N}$ and $j \in \{0, 1, \dots q + q_0\}$ in arbitrary order. Then $\{x_n\}_{n\in\mathbb{N}}$ converges to zero and $\lambda\big(\limsup_{n\to\infty}(A + x_n) \cap [0, 1]\big) = 1$. □

Proof of the theorem Observe that if $\lim_{n\to\infty} f = f$ almost everywhere, then $\lambda\big(\limsup_{n\to\infty}(A + x_n)\big) = \lambda(A)$. From the assumptions concerning A it follows that $\lambda(A) < 1$ – a contradiction.

Remark 9.5 In the proof of the lemma we used only the assumption that $\lambda(A \cap (a, b)) > 0$ for each interval $(a, b) \subset [0, 1]$ obtaining a special sequence $\{x_n\}_{n\in\mathbb{N}}$ convergent to zero for which $\lambda\big(\limsup_{n\to\infty}(A + x_n) \cap [0, 1]\big) = 1$.

The set A seems to be rather "big" and one can suppose that for each sequence $\{x_n\}_{n\in\mathbb{N}}$ the upper limit must be of the full measure. This is not the case. We shall construct two sets $A, B \subset [0, 1]$ of the first category and a sequence $\{x_n\}_{n\in\mathbb{N}}$ convergent to zero such that $\lambda(A \cap (a, b)) > 0$ and $\lambda(B \cap (a, b)) > 0$ for each interval $(a, b) \subset [0, 1]$ and $B \cap \bigcup_{n\in\mathbb{N}\cup\{0\}} (A + x_n) = \emptyset$, where $x_0 = 0$.

Let $\{(a_n, b_n)\}_{n\in\mathbb{N}}$ be a basis for the natural topology in $(0, 1)$. Choose perfect nowhere dense sets $A_1, B_1 \subset (a_1, b_1)$ and a number x_1 such that $|x_1| < 1$, $\lambda(A_1) > 0$, $\lambda(B_1) > 0$, $A_1 \cap B_1 = \emptyset$ and $(A_1 + x_1) \cap B_1 = \emptyset$. Suppose that we have chosen perfect nowhere dense sets $A_1, \dots, A_{n-1}$ and $B_1, \dots, B_{n-1}$ such that $\lambda(A_i) > 0$,

$\lambda(B_i) > 0$, $A_i, B_i \subset (a_i, b_i)$ for $i \in \{1, 2, \dots, n-1\}$ and numbers $x_1, \dots x_{n-1}$ such that $|x_i| < \frac{1}{i}$ for $i \in \{1, 2, \dots, n-1\}$ and

$$\left(\bigcup_{i=1}^{n-1} B_i\right) \cap \left(\bigcup_{j=0}^{n-1}\bigcup_{i=1}^{n-1}(A_i + x_j)\right) = \emptyset.$$

We shall construct A_n, B_n and x_n. First we choose perfect nowhere dense A_n in such a way that $A_n \subset (a_n, b_n)$, $\lambda(A_n) > 0$ and $A_n \cap \left(\bigcup_{j=0}^{n-1}\bigcup_{i=1}^{n-1}(B_i - x_j)\right) = \emptyset$. It is possible because the set $\bigcup_{j=0}^{n-1}\bigcup_{i=1}^{n-1}(B_i - x_j)$ is nowhere dense, so

$$\lambda\left((a_n, b_n) \cap \left(\bigcup_{j=0}^{n-1}\bigcup_{i=1}^{n-1}(B_i - x_j)\right)\right) < b_n - a_n.$$

We have $\left(\bigcup_{i=1}^{n-1} B_i\right) \cap \left(\bigcup_{j=0}^{n-1}\bigcup_{i=1}^{n}(A_i + x_j)\right) = \emptyset$. The set $\bigcup_{j=0}^{n-1}\bigcup_{i=1}^{n}(A_i + x_j)$ is perfect and nowhere dense, so it is possible to choose $B_n \subset (a_n, b_n)$—a perfect nowhere dense set such that $B_n \cap \left(\bigcup_{j=0}^{n-1}\bigcup_{i=1}^{n}(A_i + x_j)\right) = \emptyset$ and $\lambda(B_n) > 0$. Now we have already $\left(\bigcup_{i=1}^{n} B_i\right) \cap \left(\bigcup_{j=0}^{n-1}\bigcup_{i=1}^{n}(A_i + x_j)\right) = \emptyset$. Since the sets $\bigcup_{i=1}^{n} A_i$ and $\bigcup_{i=1}^{n} B_i$ are bounded and closed, for sufficiently small x we have $\left(\bigcup_{i=1}^{n} B_i\right) \cap \left(\bigcup_{i=1}^{n}(A_i + x)\right) = \emptyset$. We can choose x_n with this property that $|x_n| < \frac{1}{n}$. At last $\left(\bigcup_{i=1}^{n} B_i\right) \cap \left(\bigcup_{j=0}^{n}\bigcup_{i=1}^{n}(A_i + x_j)\right) = \emptyset$. So by induction we have constructed sequences $\{A_n\}_{n\in\mathbb{N}}$ and $\{B_n\}_{n\in\mathbb{N}}$ of perfect nowhere dense sets and a sequence $\{x_n\}_{n\in\mathbb{N}}$ convergent to zero such that $\left(\bigcup_{n=1}^{\infty} B_n\right) \cap \left(\bigcup_{j=0}^{\infty}\bigcup_{n=1}^{\infty}(A_n + x_j)\right) = \emptyset$. Put $A = \bigcup_{n=1}^{\infty} A_n$, $B = \bigcup_{n=1}^{\infty} B_n$. We have $B \cap \bigcup_{j=0}^{\infty}(A + x_j) = \emptyset$, so obviously

$$\limsup_{j\to\infty}(A + x_j) \subset \bigcup_{j=0}^{\infty}(A + x_j) \text{ and } \lambda\left(\limsup_{j\to\infty}(A + x_j)\right) \leq 1 - \lambda(B) < 1.$$

□

Remark 9.6 Borwein and Ditor [BD78] have proved among others that if $A \subset \mathbb{R}$ is a measurable set with $\lambda(A) > 0$ and $\{x_n\}_{n\in\mathbb{N}}$ is a sequence convergent to zero, then, for almost all $x \in A$, $x + x_n \in A$ for infinitely many n. Miller [Mi85] has shown that if $A \subset \mathbb{R}$ has the Baire property and $\{x_n\}_{n\in\mathbb{N}}$ is a sequence convergent to zero, then

$$A \setminus \{x \in A \colon x + x_n \in A \text{ for all except finitely many } n\} \in \mathcal{K}.$$

The first result is a consequence of Theorem 9.1 and the second follows immediately from Theorem 9.2. Below we shall present the result from [BD78] which can be considered as a consequence of Theorem 9.3

In the paper [RW94] a generalization of Theorem 9.1 is considered. Namely, the sequence of translations is replaced by the sequence $\{f_n\}_{n\in\mathbb{N}}$ of continuous strictly increasing functions defined on $[0, 1]$ uniformly convergent to the identity function. In [Mi89] one can find a generalization of theorems of Borwein and Ditor where the translation is replaced by the function of two variables fulfilling some natural conditions.

Theorem 9.7 ([BD78]) *There exist a measurable set $A \subset [0, 1]$ with $\lambda(A) > 0$ and a monotone sequence $\{x_n\}_{n\in\mathbb{N}}$ of positive numbers convergent to zero, such that for each $x \in A$, $x + x_n \notin A$ for infinitely many n*

Proof Let $B_1 = \{1\}$, $B_2 = \{2, 3\}$, $B_3 = \{4, 5, 6\}$ and, in general,

$$B_n = \left\{ \frac{n(n+1)}{2} + 1 - n, \frac{n(n+1)}{2} + 2 - n, \ldots, \frac{n(n+1)}{2} \right\}.$$

Let $\{i_j\}_{j\in\mathbb{N}}$ be a sequence of positive integers such that $i_j \in B_j$. For each $n \in \mathbb{N}$ let A_n be a set of numbers $x \in [0, 1]$ admitting a dyadic expansion $x = \sum_{i=1}^{\infty} \frac{a_i}{2^i}$ such that $a_{i_j} = 0$ for $j \in \{1, 2, \ldots, n\}$. Then $\{A_n\}_{n\in\mathbb{N}}$ is a descending sequence of closed sets and $\lambda(A_n) = \prod_{j=1}^{n}(1 - 2^{-j})$. Put $A = \bigcap_{k=1}^{\infty} A_k$. The set A is closed and $\lambda(A) = \prod_{j=1}^{\infty}(1 - 2^{-j}) > 0$. Let D_n be the set of all numbers x admitting a dyadic expansion $x = \sum_{i=\in B_n} \frac{a_i}{2^i}$, $a_i \in \{0, 1\}$. The set $D = \bigcup_{k=1}^{\infty} D_n$ can be enumerated to form a decreasing sequence $\{d_k\}_{k\in\mathbb{N}}$ of positive numbers convergent to zero. Suppose that $x \in A$. If n_0 is an arbitrary positive integer, then there exists $d_n \in D_{n_0}$ such that $x + d_n = \sum_{i=1}^{\infty} \frac{a_i}{2^i}$ where $a_i = 1$ for each $i \in B_{n_0}$. Hence $x + d_n \notin A$ for infinitely many n. If $x \notin A$, then $x + d_n \notin A$ for sufficiently large n because A is a closed set. □

In American Mathematical Monthly one can find the following "advanced problem 5836" ([1972]): Let f be bounded and measurable on $(0, 1)$. Is it true that $\lim_{n\to\infty} f(x - \frac{1}{n}) = f(x)$ almost everywhere? The problem has been solved by A. Meir.

Theorem 9.8 ([Mei73]) *For each $\varepsilon > 0$ there exists a bounded measurable function $f\colon (0, 1) \to \mathbb{R}$ such that*

$$\lambda\left(\left\{x \in (0, 1)\colon \limsup_{n\to\infty} \left| f\left(x - \frac{1}{n}\right) - f(x)\right| = 1\right\}\right) > 1 - \varepsilon.$$

Proof Observe that if $x \in (0, 1)$, then x can be written in the form $x = \sum_{n=2}^{\infty} \frac{a_n}{n!}$, $0 \le a_n \le n - 1$, where $a_n = [n!x] - n[(n-1)!x]$ ([] is, as usual, the integer part). Indeed, $a_2 = 0$ if $x \in (0, \frac{1}{2})$ and $a_2 = 1$ if $x \in [\frac{1}{2}, 1)$. Suppose that $a_2, a_3, \ldots, a_{n-1}$ are such that

$$\sum_{i=2}^{n-1} \frac{a_i}{i!} \le x < \sum_{i=2}^{n-1} \frac{a_i}{i!} + \frac{1}{(n-1)!}.$$

Then

$$(n-1)! \sum_{i=2}^{n-1} \frac{a_i}{i!} \le (n-1)!x < (n-1)! \sum_{i=2}^{n-1} \frac{a_i}{i!} + 1,$$

so $(n-1)! \sum_{i=2}^{n-1} \frac{a_i}{i!} = [(n-1)!x]$ and $n\left((n-1)! \sum_{i=2}^{n-1} \frac{a_i}{i!}\right) = n[(n-1)!x]$. Further

$$n! \sum_{i=2}^{n-1} \frac{a_i}{i!} \le (n-1)!x < n! \sum_{i=2}^{n-1} \frac{a_i}{i!} + n!,$$

hence $a_n = [n!x] - n! \sum_{i=2}^{n-1} \frac{a_i}{i!} = [n!x] - n[(n-1)!x]$. Let n_0 be a positive integer such that $2^{-n_0} < \varepsilon$ and lest $n_k = 2^{n_0+k}$ for $k \in \mathbb{N}$. If $B_k = \{x \in (0,1) : a_{n_k} = 0\}$, then B_k is measurable and $\lambda(B_k) = 2^{-n_0-k}$ for each $k \in \mathbb{N}$. Put $A = \bigcap_{k=1}^{\infty} \left((0,1) \setminus B_k\right)$. Then $A \in \mathcal{L}$ and $\lambda(A) > 1 - \sum_{k=1}^{\infty} 2^{-n_0-k} > 1 - \varepsilon$. Put $f = \chi_{(0,1)\setminus A}$. If $x_0 \in A$, then $x_0 = \sum_{n=2}^{\infty} \frac{a_{n_k}}{n!}$, where $a_{n_k} \neq 0$ for each $k \in \mathbb{N}$. Observe that $\frac{a_{n_k}}{n_k!} = \frac{1}{m_k}$, where m_k is a positive integer since $a_{n_k} \le n_k - 1$. The exact value of m_k is not important, all we need to know is that $\lim_{k\to\infty} m_k = \infty$. Hence $x_0 - \frac{a_{n_k}}{n_k!} = x_0 - \frac{1}{m_k} \in B_k \subset (0,1) \setminus A$ for each $k \in \mathbb{N}$ and $\lim_{k\to\infty} f\left(x_0 - \frac{1}{m_k}\right) = 1$ while $f(x_0) = 0$. The result follows from the arbitrariness of $x_0 \in A$. □

Remark 9.9 If we put $A = (0,1) \setminus \limsup B_k = (0,1) \setminus \bigcap_{pi=1}^{\infty} \bigcup_{k=pi=1}^{\infty} B_k$ and $f = \chi_{(0,1)\setminus A}$, then similarly as in the original proof in [Mei73] we obtain

$$\lambda\left(\left\{x \in (0,1) : \limsup_{n\to\infty} \left| f\left(x - \frac{1}{n}\right) - f(x)\right| = 1\right\}\right) = 1.$$

The proof of Theorem 9.8 works only for the sequence $\{\frac{1}{n}\}_{n\in\mathbb{N}}$. The situation in the case of category is quite different.

Theorem 9.10 *If $f : [0,1] \to \mathbb{R}$ has the Baire property, then for each sequence $\{a_n\}_{n\in\mathbb{N}}$ convergent to zero $\lim_{n\to\infty} f(x - a_n) = f(x)$ except on a set of the first category.*

Proof There exists a set $A \subset [0,1]$ residual on $[0,1]$ for which $f|_A$ is continuous. Put $E = A \cap \bigcap_{n=1}^{\infty}(A + a_n)$. Obviously E is also residual in $[0,1]$. If $x \in E$, then $x \in A$ and $x - a_n \in A$ for each $n \in \mathbb{N}$, so $\lim_{n\to\infty} f(x - a_n) = f(x)$. □

Definition 9.11 We say that a sequence $\{B_n\}_{n\in\mathbb{N}}$ of sets in $\mathcal{B}$ converges in category to $B \in \mathcal{B}$ ($B_n \underset{n\to\infty}{\longrightarrow} B$ in category) if and only if the sequence $\{\chi_{B_n}\}_{n\in\mathbb{N}}$ of characteristic functions converges in category to χ_B. Similarly we say that a sequence $\{B_n\}_{n\in\mathbb{N}}$ of measurable sets converges in measure to $B \in \mathcal{L}$ ($B_n \underset{n\to\infty}{\longrightarrow} B$ in measure) if and only if the sequence $\{\chi_{B_n}\}_{n\in\mathbb{N}}$ converges in measure to χ_B.

Remark 9.12 Since in both cases $\{B_n\}_{n\in\mathbb{N}}$ converges to B if and only if $\{B_n \triangle B\}_{n\in\mathbb{N}}$ converges to the empty set, in the sequel we shall consider mainly the convergence to $\emptyset$. Obviously $A_n \underset{n\to\infty}{\longrightarrow} \emptyset$ in measure if and only if $\lambda(A_n) \underset{n\to\infty}{\longrightarrow} 0$.

The following theorems follow immediately from the definition and from the fact that if measurable functions (defined on $[0, 1]$) in the sequence are commonly bounded, then convergence in measure is equivalent to convergence in L^1.

Theorem 9.13 *If $\{A_n\}_{n\in\mathbb{N}}$ is a sequence of measurable subsets of $[0, 1]$ such that $A_n \underset{n\to\infty}{\longrightarrow} \emptyset$ in measure, then the sequence $f_n = \frac{1}{n}\sum_{i=1}^{n} \chi_{A_i}$ converges to 0 in measure. If a sequence $\{\chi_{A_n}\}_{n\in\mathbb{N}}$ converges to zero almost everywhere, then the sequence $\{f_n\}_{n\in\mathbb{N}}$ also converges to zero almost everywhere.*

Theorem 9.14 *If $\{A_n\}_{n\in\mathbb{N}}$ is a sequence of measurable subsets of $[0, 1]$ such that $A_n \underset{n\to\infty}{\longrightarrow} \emptyset$ in measure and $\{r_n\}_{n\in\mathbb{N}}$ is a sequence of real numbers convergent to zero, then $(A_n + r_n) \cap [0, 1] \underset{n\to\infty}{\longrightarrow} \emptyset$ in measure.*

Observe that if $\{A_n\}_{n\in\mathbb{N}}$ is a sequence of measurable subsets of $[0, 1]$, then the sequence $\{\chi_{A_n}\}_{n\in\mathbb{N}}$ converges almost everywhere to 0 if and only if $\lambda(\limsup_{n\to\infty} A_n) = 0$. Below we shall study the behaviour of $\lambda\big(\limsup_{n\to\infty}(A_n + \varepsilon_n)\big)$ when $\lim_{n\to\infty} \varepsilon_n = 0$ under the assumption $\lambda(\limsup_{n\to\infty} A_n) = 0$ (see [WW05]).

Theorem 9.15 *For each sequence $\{A_n\}_{n\in\mathbb{N}}$ of measurable subsets of $[0, 1]$ such that $\lambda(\limsup_{n\to\infty} A_n) = 0$ there exists a sequence $\{\varepsilon_n\}_{n\in\mathbb{N}}$ of numbers different from zero such that $\lim_{n\to\infty} \varepsilon_n = 0$ and $\lambda\big(\limsup_{n\to\infty}(A_n + \varepsilon_n)\big) = 0$.*

Proof For each $n \in \mathbb{N}$ there exists $\varepsilon_n \in (0, \frac{1}{n})$ such that $\lambda\big(A_n \triangle (A_n + \varepsilon_n)\big) < \frac{1}{2^n}$. Observe that $(A_m + \varepsilon_m) \subset A_m \cup \big((A_m + \varepsilon_m)\triangle A_m\big)$ for each $m \in \mathbb{N}$, so

$$\bigcup_{m=n}^{\infty}(A_m + \varepsilon_m) \subset \bigcup_{m=n}^{\infty} A_m \cup \bigcup_{m=n}^{\infty} \big((A_m + \varepsilon_m)\triangle A_m\big)$$

and

$$\lambda\Big(\bigcup_{m=n}^{\infty}(A_m + \varepsilon_m)\Big) \le \lambda\Big(\bigcup_{m=n}^{\infty} A_m\Big) + 2^{1-n}.$$

Since $\lambda\big(\bigcup_{m=n}^{\infty} A_m\big) = 0$, we obtain that $\lim_{n\to\infty} \lambda\big(\bigcup_{m=n}^{\infty}(A_m + \varepsilon_m)\big) = 0$ which ends the proof. □

Remark 9.16 If $\sum_{n=1}^{\infty} \lambda(A_n) < \infty$, then obviously $\lambda\big(\limsup_{n\to\infty}(A_n + \varepsilon_n)\big) = 0$ for arbitrary sequence $\{\varepsilon_n\}_{n\in\mathbb{N}}$.

However, if $\sum_{n=1}^{\infty} \lambda(A_n) = \infty$ the situation is different. For example, if $A_n = [0, \frac{1}{n}]$ for $n \in \mathbb{N}$, then $\lambda(\limsup_{n\to\infty} A_n) = 0$ and also $\lambda\big(\limsup_{n\to\infty}(A_n + \varepsilon_n)\big) = 0$ for each sequence $\{\varepsilon_n\}_{n\in\mathbb{N}}$ convergent to zero (but not for arbitrary sequence $\{\varepsilon_n\}_{n\in\mathbb{N}}$). Put $a_k = \frac{(k-1)k}{2}$ and $E_n = \bigcup_{i=0}^{2^{a_k}-1}\left(\frac{i}{2^{a_k}}, \frac{i}{2^{a_k}} + \frac{1}{2^{a_k+1}}\right)$ for $k \in \mathbb{N}$. Now let $A_n = E_k$ if $\sum_{i=1}^{k-1} 2^i + 1 \le n \le \sum_{i=1}^{k} 2^i$ and $\varepsilon_n = \frac{j-1}{2^{a_k+1}}$ if $n = \sum_{i=1}^{k-1} 2^i + j$, where $1 \le j \le 2^k$. The sequence $\{\varepsilon_n\}_{n\in\mathbb{N}}$ converges to zero. Moreover, $\lambda(E_k) = 2^{-k}$ for $k \in \mathbb{N}$, so $\sum_{n=1}^{\infty} \lambda(A_n) = \infty$. Observe that $\limsup_{n\to\infty} A_n = \limsup_{n\to\infty} E_k$, so $\lambda\big(\limsup_{n\to\infty} A_n\big) = 0$. At the same time $\lambda\big(\bigcup_{m=m_k}^{M_k}(A_n + \varepsilon_n)\big) = 1$ for each $k \in \mathbb{N}$, where $m_k = 1 + \sum_{i=1}^{k-1} 2^i$ and $M_k = \sum_{i=1}^{2^k} 2^i$, hence $\lambda(\limsup_{n\to\infty}\big(A_n + \varepsilon_n)\big) = 1$.

In the case of category the situation is simpler.

Theorem 9.17 *For each sequence $\{A_n\}_{n\in\mathbb{N}}$ of sets having the Baire property such that* $\limsup_{n\to\infty} A_n \in \mathcal{K}$ *and for each sequence $\{\varepsilon_n\}_{n\in\mathbb{N}}$ convergent to zero we have* $\lambda\big(\limsup_{n\to\infty}(A_n + \varepsilon_n)\big) \in \mathcal{K}$.

Proof Since $A_n = G_n \triangle K_n$, where G_n is open and $K_n \in \mathcal{K}$, we have $\limsup_{n\to\infty} A_n \triangle \limsup_{n\to\infty} G_n \in \mathcal{K}$, we shall consider $\limsup_{n\to\infty} G_n$ and $\limsup_{n\to\infty}(G_n + \varepsilon_n)$. Observe that $\limsup_{n\to\infty} G_n = \bigcap_{m=1}^{\infty}\bigcup_{n=m}^{\infty} G_n$ as G_δ set of the first category must be nowhere dense. We shall show that $\limsup_{n\to\infty}(G_n + \varepsilon_n)$ is also nowhere dense, which is obviously will suffice. Take an arbitrary interval $(a, b) \subset [0, 1]$. There exists $m_0 \in \mathbb{N}$ such that $\bigcup_{m=m_0}^{\infty} G_n$ is not dense in (a, b) (in the contrary case $(a, b) \cap \limsup_{n\to\infty} G_n$ would be of the second category), which means that there exists an interval $(c_0, d_0) \subset (a, b)$ such that $(c_0, d_0) \cap \bigcup_{n=m_0}^{\infty} G_n = \emptyset$. Let now $m \ge m_0$ be a positive integer such that $|\varepsilon_n| < \frac{1}{3}(d_0 - c_0)$ for $n \ge m_1$. If $(c, d) = \big(c_0 + \frac{1}{3}(d_0 - c_0), d_0 - \frac{1}{3}(d_0 - c_0)\big)$, then $(c, d) \cap (G_n + \varepsilon_n) = \emptyset$ for $n \ge m_1$, which implies $(c, d) \cap \bigcup_{m=m_1}^{\infty}(G_n + \varepsilon_n) = \emptyset$ and $(c, d) \cap \limsup_{n\to\infty}(G_n + \varepsilon_n) = \emptyset$. Since $(c, d) \subset (a, b)$ and (a, b) was arbitrary, $\limsup_{n\to\infty}(G_n + \varepsilon_n)$ is nowhere dense and $\limsup_{n\to\infty}(A_n + \varepsilon_n) \in \mathcal{K}$. □

Theorem 9.18 *If $\{A_n\}_{n\in\mathbb{N}}$ is a sequence of measurable subsets of* $[0, 1]$ *such that the sequence $f_n = \frac{1}{n}\sum_{i=1}^{n} \chi_{A_i}$, $n \in \mathbb{N}$, converges in measure to* 0, *and $\{r_n\}_{n\in\mathbb{N}}$ is a sequence of real numbers convergent to* 0, *then the sequence $g_n = \frac{1}{n}\sum_{i=1}^{n} \chi_{B_i}$, where $B_n = (A_n + r_n) \cap [0, 1]$, $n \in \mathbb{N}$ also converges in measure to* 0.

To study the convergence in category we shall need the following lemma in which $\tilde{B}_n$ is the regular open set such that $B_n \triangle \tilde{B}_n \in \mathcal{K}$:

Lemma 9.19 ([Wil16]) *Let $\{B_n\}_{n\in\mathbb{N}}$ be a sequence of subsets of* $[0, 1]$ *having the Baire property. The following conditions are equivalent:*

(1) $B_n \underset{n\to\infty}{\longrightarrow} \emptyset$ *in category,*
(2) for every nonempty interval $[a, b] \subset [0, 1]$ there exist a number $\varepsilon > 0$ and a positive integer n_0 such that for every $n \ge n_0$ there exists an interval $(c_n, d_n) \subset [a, b]$ with the properties $|d_n - c_n| > \varepsilon$ and $(c_n, d_n) \cap \tilde{B}_n = \emptyset$,

(3) *for every nonempty interval of the form* $\left[\frac{p-1}{2^k}, \frac{p}{2^k}\right]$, *where* $k \in \mathbb{N}$ *and* $p \in \{1, 2, \ldots, 2^k\}$ *there exist a number* $\varepsilon > 0$ *and a positive integer* n_0 *such that for every* $n \geq n_0$ *there exists an interval* $(c_n, d_n) \subset \left[\frac{p-1}{2^k}, \frac{p}{2^k}\right]$ *with the properties that* $|d_n - c_n| > \varepsilon$ *and* $(c_n, d_n) \cap \tilde{B}_n = \emptyset$.

Proof (1) $\Longrightarrow$ (2). Suppose that (2) does not hold. There exists a closed interval $I \subset [0, 1]$ such that for every positive integer $k > \frac{1}{\lambda(I)}$ there exist an increasing sequence $\{n^k_m\}_{m\in\mathbb{N}}$ of positive integers such that for each open interval $J \subset I$ such that $\lambda(J) > \frac{1}{k}$ we have $J \cap \tilde{B}_{n^k_m} \neq \emptyset$ for all $m \in \mathbb{N}$. Choose a sequence $\{m_k\}_{k>\frac{1}{\lambda(I)}}$ such that $\{n^k_{m_k}\}_{k>\frac{1}{\lambda(I)}}$ is an increasing sequence. If $n_k = n^k_{m_k}$ for $k > \frac{1}{\lambda(I)}$, then for each subsequence $\{n_{k_p}\}_{p\in\mathbb{N}}$ of $\{n_k\}_{k>\frac{1}{\lambda(I)}}$ and for each $q \in \mathbb{N}$ the set $(\bigcup_{p\geq q} \tilde{B}_{n_{k_p}}) \cap I$ is dense and open in I, so 1) does not hold.

(2) $\Longrightarrow$ (3). This is obvious.

(3) $\Longrightarrow$ (1). Let $I_{2^k+p-1} = \left[\frac{p-1}{2^k}, \frac{p}{2^k}\right]$, for $k \in \mathbb{N} \cup \{0\}$, $p \in \{1, 2, \ldots, 2^k\}$.

Take an increasing sequence $\{n_m\}_{m\in\mathbb{N}}$ of positive integers. We shall construct a subsequence $\{n_{m_p}\}_{m\in\mathbb{N}}$ of $\{n_m\}_{m\in\mathbb{N}}$ for which $\chi_{\tilde{B}_{n_{m_p}}} \underset{p\to\infty}{\longrightarrow} \chi_\emptyset$ except on the set of the first category, which yields $B_n \underset{n\to\infty}{\longrightarrow} B$ in category.

From the assumption it follows that there exist $\varepsilon_1 > 0$ and a number $m_0 \in \mathbb{N}$ such that for $m \geq m_0$ there exists an open interval $(c_m, d_m) \subset I_1$ such that $d_m - c_m > \varepsilon_1$ and $(c_m, d_m) \cap \tilde{B}_m = \emptyset$. Choose a subsequence $\{n_{m_j}\}_{j\in\mathbb{N}}$ for which there exists $\lim_{j\to\infty} c_{m_j} = c^1$. Then we have $\lim_{j\to\infty} \chi_{\tilde{B}_{n_{m_j}}}(x) = 0$ for $x \in J_1 = (c^1, c^1 + \varepsilon_1) \subset I_1$.

For convenience we shall denote the sequence $\{n_{m_j}\}_{j\in\mathbb{N}}$ by $\{n^1_m\}_{m\in\mathbb{N}}$.

Similarly for I_2 we choose a subsequence $\{n^2_m\}_{m\in\mathbb{N}}$ of $\{n^1_m\}_{m\in\mathbb{N}}$ for which there exists a nondegenerated open interval $J_2 \subset I_2$ such that $\lim_{m\to\infty} \chi_{\tilde{B}_{n^2_m}}(x) = 0$ for $x \in J_2$.

By induction we define a sequence of subsequences $\{\{n^i_m\}_{m\in\mathbb{N}}\}_{i\in\mathbb{N}}$ such that $\{n^{i+1}_m\}_{m\in\mathbb{N}}$ is a subsequence of $\{n^i_m\}_{m\in\mathbb{N}}$ and $\lim_{m\to\infty} \chi_{\tilde{B}_{n^i_m}}(x) = 0$ for $x \in J_i$, where $J_i \subset I_i$ is a nondegenerated open interval.

If $\{n_{m_p}\}_{p\in\mathbb{N}}$ is a diagonal sequence for $\{\{n^i_m\}_{m\in\mathbb{N}}\}_{i\in\mathbb{N}}$, then $\lim_{p\to\infty} \chi_{\tilde{B}_{n_{m_p}}}(x) = 0$ for $x \in \bigcup_{i=1}^{\infty} J_i$. Since $\bigcup_{i=1}^{\infty} J_i$ is an open set dense in $[0, 1]$, we even obtain convergence to zero off a nowhere dense set. □

Now we shall show that only the analogues of Theorem 9.15 and the second part of Theorem 9.13 are true in the category case.

Theorem 9.20 ([Wil16]) *If* $\{B_n\}_{n\in\mathbb{N}}$ *is a sequence of subsets of* $[0, 1]$ *having the Baire property such that* $B_n \underset{n\to\infty}{\longrightarrow} \emptyset$ *in category and* $\{r_n\}_{n\in\mathbb{N}}$ *is a sequence of real numbers convergent to zero, then* $(B_n + r_n) \cap [0, 1] \underset{n\to\infty}{\longrightarrow} \emptyset$ *in category.*

Proof Take $[a, b] \subset [0, 1]$, $[a, b]$—a nondegenerated interval. From the condition (2) of the lemma it follows that there exist $\varepsilon > 0$ and $n_1 \in \mathbb{N}$ such that for each

$n \geq n_1$ there exists an interval $(c_n, d_n) \subset [a, b]$ with the properties that $d_n - c_n > \varepsilon$ and $(c_n, d_n) \cap \tilde{B}_n = \emptyset$. Since $r_n \underset{n\to\infty}{\longrightarrow} 0$, there exists $n_2 \in \mathbb{N}$ such that $|r_n| < \frac{\varepsilon}{2}$ for $n \geq n_2$. Put $(e_n, f_n) = (c_n + r_n, d_n)$ if $r_n > 0$ and $(e_n, f_n) = (c_n, d_n + r_n)$ if $r_n < 0$. Then we have $(e_n, f_n) \subset [a, b] \cap (c_n + r_n, d_n + r_n)$ and $|f_n - e_n| > \frac{\varepsilon}{2}$ for $n \geq n_0 = \max(n_1, n_2)$. Since $(c_n + r_n, d_n + r_n) \cap (\tilde{B}_n + r_n) = \emptyset$, then $(e_n, f_n) \cap (\tilde{B}_n + r_n) = \emptyset$ for $n \geq n_0$. From the Lemma 9.19 it follows immediately that $(B_n + r_n) \cap [0, 1] \underset{n\to\infty}{\longrightarrow} \emptyset$ in category. □

Theorem 9.21 ([Wil16]) *There exists a sequence* $\{B_n\}_{n\in\mathbb{N}}$ *of subsets of* $[0, 1]$ *having the Baire property such that* $B_n \underset{n\to\infty}{\longrightarrow} \emptyset$ *in category but the sequence* $f_n = \frac{1}{n}\sum_{i=1}^{n} \chi_{B_i}$ *does not converge to* 0 *in category.*

Proof Put $n_0 = 0$ and $n_j = \sum_{i=1}^{j} 4^i$ for $j \in \mathbb{N}$. The definition of B_n will proceed in step-intervals of size 4^k (thus motivating the use of the numbers n_j). As a preliminary we define the sets A_k^i which are 4^k contiguous subsets of $(0, \frac{1}{2^{k-1}})$ of equal lengths. Let $A_k^i = \left(\frac{i}{4^k.2^{k-1}}, \frac{i+1}{4^k.2^{k-1}}\right)$ for $k \in \mathbb{N}$ and $i \in \{0, 1, \ldots, 4^k - 1\}$. Put $C_k^i = \bigcup_{j=0}^{2^{k-1}-1}\left(A_k^i + \frac{j}{2^{k-1}}\right)$ for $k \in \mathbb{N}$ and $i \in \{0, 1, \ldots, 4^k - 1\}$. Each C_k^i, the union of the 2^{k-1} shifts of A_k^i, has measure 4^{-k}.

Observe that for fixed $k \in \mathbb{N}$ the set C_k^i are pairwise disjoint and $[0, 1] \setminus \bigcup_{i=0}^{4^k-1} C_k^i$ is a finite set.

Put

$$B_n = [0, 1] \setminus C_1^n = [0, 1] \setminus A_1^n \quad \text{for } n \in \{0, 1, 2, 3\} = \{n_0, n_0 + 1, n_0 + 2, n_1 - 1\}$$

and

$$B_n = [0, 1] \setminus (C_1^{i_1} \cup C_2^{i_2}) \quad \text{for } n \in \{n_1, n_1 + 1, \ldots, n_2 - 1\},$$

where

$$i_1 \equiv (n - n_1) \pmod{4^1}$$

and

$$i_2 \equiv (n - n_1) \pmod{4^2}.$$

Generally, let $B_n = [0, 1] \setminus \bigcup_{m=1}^{j} C_m^{i_m}$ for $n \in \{n_{j-1}, n_{j-1} + 1, \ldots, n_j - 1\}$, where $i_m \equiv (n - n_{j-1}) (\mathrm{mod}\, 4^m)$ for $m \in \{1, 2, \ldots, j\}$. The set B_n (where $n \in \{n_{j-1}, n_{j-1} + 1, \ldots n_j - 1\}$) omits intervals with sum-length $4^{-1} + 4^{-2} + \cdots + 4^{-j}$.

Observe that the sequence $\{B_n\}_{n\in\mathbb{N}\cup\{0\}}$ of closed sets fulfills the condition 3) from the lemma. Indeed, for each interval $\left[\frac{p-1}{2^{k-1}}, \frac{p}{2^{k-1}}\right]$, and for each positive $\varepsilon < \frac{1}{4^k.2^{k-1}}$ and for each $n \geq n_{k-1}$ there exists are open subinterval of length exactly $\frac{1}{4^k.2^{k-1}}$ disjoint with $\tilde{B}_n = \mathrm{Int}\, B_n$. Hence $B_n \underset{n\to\infty}{\longrightarrow} \emptyset$ in category. Simultaneously for each $k \in \mathbb{N}$ and for each $x \in [0, 1]$ we have by reference to the extreme case

$$n_k f_{n_k}(x) = \sum_{n=1}^{n_k} \chi_{B_n}(x) \geq n_k - n_k\left(\frac{1}{4} + \frac{1}{4^2} + \cdots + \frac{1}{4^{k-1}}\right) > \frac{2}{3}n_k,$$

so $\frac{1}{n_k}\sum_{n=1}^{n_k} \chi_{B_n}(x) \geq \frac{2}{3}$ for each $x \in [0, 1]$ and each $k \in \mathbb{N}$, which means that the sequence $\{f_n\}_{n\in\mathbb{N}}$ does not converge to 0 in category. □

Theorem 9.22 ([Wil16]) *If a sequence $\{\chi_{A_n}\}_{n\in\mathbb{N}}$ converges to zero except on a set of the first category, then the sequence $\{f_n\}_{n\in\mathbb{N}}$, where $f_n = \frac{1}{n}\sum_{i=1}^{n} \chi_{A_i}$ also converges to zero except on a set of the first category.*

Proof Put $E = \{x \in [0, 1]\colon \lim_{n\to\infty} \chi_{A_n}(x) = 0\}$. Obviously $E = \liminf_{n\to\infty} A_n = \bigcup_{n=1}^{\infty}\bigcap_{m=n}^{\infty} A_n$ and E is residual in $[0, 1]$. Observe that if $x \in E$, then $\lim_{n\to\infty} \frac{1}{n}\sum_{i=1}^{n} \chi_{A_i}(x) = 0$, since there exists $n \in \mathbb{N}$ such that $\chi_{A_m}(x) = 0$ for $m \geq n$. Hence the set $\{x \in [0, 1]\colon \lim_{n\to\infty} \frac{1}{n}\sum_{i=1}^{n} \chi_{A_i}(x) = 0\}$ is also residual in $[0, 1]$. □

Theorem 9.23 ([Wil16]) *There exists a sequence $\{A_n\}_{n\in\mathbb{N}}$ of sets having the Baire property such that $\lim_{n\to\infty} \frac{1}{n}\sum_{i=1}^{n} \chi_{A_i}(x) = 0$ except on a set of the first category but $\{\frac{1}{n}\sum_{i=1}^{n} \chi_{A_i+3^{-i}}(x)\}_{n\in\mathbb{N}}$ does not converge to 0 with respect to $\mathcal{I}$.*

Proof Put $r_n = 3^{-n}$ for $n \in \mathbb{N}$ and

$$D_k = \bigcup_{i=1}^{2^{k-1}} \left(\frac{i}{2^{k-1}} - r_{2^k}, \frac{i}{2^{k-1}}\right) \quad \text{for } k \in \mathbb{N}.$$

Finally let $A_n = (D_k - r_n) \cap (0, 1)$ for $n \in \{2^{k-1}, \ldots, 2^k - 1\}$, $k \in \mathbb{N}$. Observe that for each $k \in \mathbb{N}$ the components of D_k are narrow enough that 2^{k-1} translates of A_n for $2^{k-1} \leq n < 2^k$ are disjoint. Hence for each $x \in (0, 1)$ we have $\lim_{n\to\infty} \frac{1}{n}\sum_{i=1}^{n} \chi_{A_i}(x) = 0$. At the same time $A_n + r_n = D_k$ for $n \in \{2^{k-1}, \ldots, 2^k - 1\}$ and the set $D = \limsup_{k\to\infty} D_k$ is a dense G_δ set. If $x \in D$, then for infinitely many k (referring to $n = 2^k$) we have $x \in D_k$ and then $x \in \bigcap_{n=2^{k-1}}^{2^k-1}(A_n + r_n)$, hence $\limsup_{n\to\infty} \frac{1}{n}\sum_{i=1}^{n} \chi_{A_i+r_i}(x) \geq \frac{1}{2}$. That means that $\{\frac{1}{n}\sum_{i=1}^{n} \chi_{A_i+r_i}\}_{n\in\mathbb{N}}$ does not converge to 0 except on a set of the first category. To prove that $\{\frac{1}{n}\sum_{i=1}^{n} \chi_{A_i+3^{-i}}\}_{n\in\mathbb{N}}$ does not converge to 0 in category, take an arbitrary increasing sequence $\{n_m\}_{m\in\mathbb{N}}$ of positive numbers. For each $m \in \mathbb{N}$ there exists $k_m \in \mathbb{N}$ such that $2^{k_m-1} \leq m \leq 2^{k_m} - 1$. If $n_m < 2^{k_m-1} + 2^{k_m-2}$, then $\frac{1}{n_m}\sum_{i=1}^{n_m} \chi_{A_i+r_i}(x) > \frac{1}{4}$ for $x \in D_{k_m-1}$ and if $n_m \geq 2^{k_m-1} + 2^{k_m-2}$, then $\frac{1}{n_m}\sum_{i=1}^{n_m} \chi_{A_i+r_i}(x) > \frac{1}{4}$ for $x \in D_{k_m}$. Take $p_m = k_m - 1$ in the first case and $p_m = k_m$ in the second; then the $\frac{1}{4}$ lower bound holds for $x \in \limsup_{n\to\infty} D_{p_m}$, a dense G_δ set. From the arbitrariness of $\{n_m\}_{m\in\mathbb{N}}$ the claim follows immediately. □

Chapter 10
Relative Compactness and Measurability of Functions of Accumulation

Fremlin [Fre75] has proved the following theorem: If $(X, \mathcal{S}, \mu)$ is a perfect measure space and $\{f_n\}_{n\in\mathbb{N}}$ —a sequence of real-valued $\mathcal{S}$-measurable functions defined on X, then one of the following conditions holds:

(a) there exists a subsequence $\{f_{n_k}\}_{k\in\mathbb{N}}$ convergent μ-a.e.,
(b) there exists a subsequence $\{f_{n_k}\}_{k\in\mathbb{N}}$ such that each function of accumulation of this sequence is not measurable.

Recall that $f : X \to \mathbb{R}$ is a function of accumulation of a sequence $\{f_n\}_{n\in\mathbb{N}}$ if for each $\varepsilon > 0$ and each finite sequence $\{x_1, x_2, \ldots, x_k\}$ of elements of X there exists a subsequence $\{f_{n_m}\}_{m\in\mathbb{N}}$ such that $|f(x_i) - f_{n_m}(x_i)| < \varepsilon$ for each $m \in \mathbb{N}$ and $i \in \{1, 2, \ldots, k\}$ (see [Tyc35]).

The first result in this direction was obtained by Sierpiński [Sie45] who constructed a sequence of functions belonging to the first Baire class (very similar to the Rademacher orthogonal sequence) such that each function of accumulation of this sequence is not measurable. In [Sie47] Sierpiński modified the construction to obtain the sequence of continuous functions having the same property.

The proof of Fremlin uses some properties of weak convergence in $L^2(\mu)$ (compare also [Ta-d84]), so there is no possibility to find similar reasoning for the convergence except on a set of the first category ($\mathcal{K}$-a.e.). Bucchioni and Goldman [BG78] have presented a weaker version of theorem of Fremlin with the proof using only basic properties of a measure space. Their proof can work also for real-valued functions defined on [0, 1] and for Baire category after rather small completion.

In the sequel $\mathcal{S}$ should be interpreted as $\mathcal{L}$ or $\mathcal{B}$, and $\mathcal{I}$—as $\mathcal{N}$ or $\mathcal{K}$, all reduced to the unit interval [0, 1].

Lemma 10.1 *$B \notin \mathcal{S}$ if and only if there exists a set $Y \in \mathcal{S} \setminus \mathcal{I}$ such that for each set $A \subset Y$ if $A \in \mathcal{S} \setminus \mathcal{I}$, then $A \cap B \neq \emptyset$ and $A \setminus B \neq \emptyset$.*

Proof Necessity. Suppose that it is not the case, so for each set $Y \in \mathcal{S} \setminus \mathcal{I}$ there exists a set $A \subset Y$ such that $A \in \mathcal{S} \setminus \mathcal{I}$ and $A \cap B = \emptyset$ or $A \setminus B = \emptyset$. Take $Y_0 = [0, 1]$. Then there exists a set $A_0 \subset Y_0$ such that $A_0 \subset \mathcal{S} \setminus \mathcal{I}$ and $A_0 \cap B = \emptyset$ or $A_0 \setminus B = \emptyset$.

W. Wilczyński, *Convergence in Measure and in Category*, Lecture Notes in Mathematics 2394, https://doi.org/10.1007/978-3-032-23143-7_10

Suppose that the sets A_β for $\beta < \alpha$, where α is a countable ordinal, have been defined. Put $Y_\alpha = [0, 1] \setminus \bigcup_{\beta<\alpha} A_\beta$. If $Y_\alpha \in \mathcal{I}$, the procedure stops. If not, then there exists a set $A_\alpha \subset Y_\alpha$, $A_\alpha \subset \mathcal{S} \setminus \mathcal{I}$ such that $A_\alpha \cap B = \emptyset$ or $A_\alpha \setminus B = \emptyset$. There exists the smallest ordinal number $\alpha_0 < \omega_1$ for which $Y_{\alpha_0} \in \mathcal{I}$, since the pair $(\mathcal{S}, \mathcal{I})$ fulfills the countable chain condition (both in measure and in category case). If $D = \{\beta < \alpha_0 \colon A_\beta \subset B\} = \{\beta < \alpha_0 \colon A_\beta \setminus B = \emptyset\}$, then D is a countable set and $B = \left(\bigcup_{\beta\in D} A_\beta\right) \cup B_0$, where $B_0 \subset Y_{\alpha_0}$, so $B \in \mathcal{S}$.

Sufficiency. Suppose now that B is measurable. Take arbitrary $Y \in \mathcal{S} \setminus \mathcal{I}$. If $Y \cap B \in \mathcal{I}$, then take $A = Y \setminus B$. We have $A \in \mathcal{S} \setminus \mathcal{I}$ and $A \cap B = \emptyset$. If $Y \cap B \in \mathcal{S} \setminus \mathcal{I}$, take $A = Y \cap B$. □

Lemma 10.2 *If $f \colon [0, 1] \to \mathbb{R}$ is not $\mathcal{S}$-measurable, then there exist a set $Y \in \mathcal{S} \setminus \mathcal{I}$ and two numbers: $r_0 \in \mathbb{R}$ and $\delta_0 > 0$ such that for each set $A \subset Y$ if $A \in \mathcal{S} \setminus \mathcal{I}$, then there exist $x_1, x_2 \in A$ such that $f(x_1) > r_0 + \delta_0$ and $f(x_2) < r_0$.*

Proof There exists $r \in \mathbb{R}$ such that $f^{-1}((r, \infty)) \notin \mathcal{S}$. From Lemma 10.1 it follows that there exists a set $Z \in \mathcal{S} \setminus \mathcal{I}$ such that for each set $A \subset Z$ if $A \in \mathcal{S} \setminus \mathcal{I}$, then there exist $x_1, x_2 \in A$ such that $f(x_1) > r$ and $f(x_2) \le r$. We shall show that there exist a set $Y \subset Z$, $Y \in \mathcal{S} \setminus \mathcal{I}$ and a number $\delta > 0$ such that if $A \subset Y$, $A \in \mathcal{S} \setminus \mathcal{I}$, then there exist $x_1, x_2 \in A$ such that $f(x_1) > r + \delta$ and $f(x_2) \le r$. Suppose that this is not the case. That means that for each set $Y \subset Z$, $Y \in \mathcal{S} \setminus \mathcal{I}$ and each number $\delta > 0$ there exists a set $A \subset Y$, $A \in \mathcal{S} \setminus \mathcal{I}$ for which $A \subset f^{-1}((-\infty, r + \delta])$ or $A \subset f^{-1}((r, \infty))$. The last inclusion does not hold, because $A \subset Y \subset Z$.

Fix $\delta > 0$ for a moment. Put $Y_0 = Z$. There exists $A_0 \subset Y_0$, $A_0 \in \mathcal{S} \setminus \mathcal{I}$ for which $A_0 \subset f^{-1}((-\infty, r + \delta])$. Suppose that we have constructed Y_β and A_β for $\beta < \alpha$ (where α is a countable ordinal). Take $Y_\alpha = Z \setminus \bigcup\limits_{\beta<\alpha} A_\beta$. If $Y_\alpha \in \mathcal{I}$, the procedure stops. If $Y_\alpha \in \mathcal{S} \setminus \mathcal{I}$, then exists $A_\alpha \in \mathcal{S} \setminus \mathcal{I}$ for which $A_\alpha \subset f^{-1}((-\infty, r + \delta])$. There exists $\alpha_0 < \omega_1$ for which $Y_{\alpha_0} \in \mathcal{I}$, since $(\mathcal{S}, \mathcal{I})$ fulfills (ccc). Hence $f(x) \le r + \delta$ $\mathcal{I}$-a.e. on Z. From the arbitrariness of $\delta > 0$ it follows that $f(x) \le r$ $\mathcal{I}$-a.e. on Z, which contradicts the property of Z. So there exist $Y \subset Z$, $Y \in \mathcal{S} \setminus \mathcal{I}$ and $\delta > 0$ such that for each $A \subset Y$, $A \in \mathcal{S} \setminus \mathcal{I}$ there exist $x_1, x_2 \in A$ such that $f(x_1) > r + \delta$ and $f(x_2) \le r$. To finish the proof it is enough to take $r_0 \in (r, r + \delta)$ and δ_0 such that $r_0 + \delta_0 = r + \delta$. □

Lemma 10.3 *If $f = \chi_B$, where $B \in \mathcal{S}$, then for each set $Y \in \mathcal{S} \setminus \mathcal{I}$ and for each numbers $r \in \mathbb{R}$ and $\delta > 0$ there exists a set $A \subset Y$, $A \in \mathcal{S} \setminus \mathcal{I}$ such that $f(A) \subset (-\infty, r + \delta]$ or $f(A) \subset [r, \infty)$.*

Proof Take $Y \in \mathcal{S} \setminus \mathcal{I}$. If $Y \cap B \in \mathcal{I}$, then $Y \setminus B \in \mathcal{S} \setminus \mathcal{I}$, so for $r \le 0$ and arbitrary $\delta > 0$ it is sufficient to take $A = Y \setminus B$ and then $f(A) \subset [r, \infty)$; for $r > 0$ and arbitrary $\delta > 0$ take also $A = Y \setminus B$ and we have $f(A) \subset (-\infty, r + \delta]$. If $Y \cap B \in \mathcal{S} \setminus \mathcal{I}$, take $A = Y \cap B$. Then for $r \le 1$ and arbitrary $\delta > 0$ we have $f(A) \subset [r, \infty)$ and for $r > 1$ and arbitrary $\delta > 0$ we have $f(A) \subset (-\infty, r + \delta]$. □

The above lemma shows that the condition from Lemma 10.2 is not only necessary for a function to be not $\mathcal{S}$-measurable, but also sufficient if a function is a characteristic function of a set. Below we shall prove that it is also sufficient for arbitrary function. Modifying the proof of Lemma 10.3 one can show that it is true for simple functions of the form $f = \sum_{k=1}^{n} a_k \cdot \chi_{B_k}$.

Theorem 10.4 *A function $f\colon [0,1] \to \mathbb{R}$ is not $\mathcal{S}$-measurable if and only if there exist a set $Y \in \mathcal{S} \setminus \mathcal{I}$ and two numbers: $r \in \mathbb{R}$ and $\delta > 0$ such that for each set $A \subset Y$, $A \in \mathcal{S} \setminus \mathcal{I}$ there exist $x_1, x_2 \in A$ such that $f(x_1) > r + \delta$ and $f(x_2) < r$.*

Proof We must prove only sufficiency. Suppose that the condition is fulfilled for a function f.

(a) Measure case. Let $Y \in \mathcal{L} \setminus \mathcal{N}$, $r \in \mathbb{R}$ and $\delta > 0$ be a set and numbers whose existence is guaranteed in Lemma 10.2. If f is Lebesgue measurable, then there exists a sequence $\{f_n\}_{n\in\mathbb{N}}$ of simple measurable functions convergent a.e. to f. Take $\varepsilon \in (0, \lambda(Y))$. From the theorem of Egorov it follows that there exist a measurable set Y_ε such that $\lambda(Y \setminus Y_\varepsilon) \leq \frac{\varepsilon}{2}$ and a positive integer n_0 such that $\sup\{|f(x) - f_{n_0}(x)|\colon x \in Y_\varepsilon\} < \frac{\delta}{4}$. Observe that $\lambda(Y \cap Y_\varepsilon) > \frac{\varepsilon}{2} > 0$. Take an arbitrary measurable set $A \subset Y \cap Y_\varepsilon$, $\lambda(A) > 0$. There exist $x_1, x_2 \in A$ such that $f(x_1) > r + \delta$ and $f(x_2) < r$. But then $f_{n_0}(x_1) > r + \frac{3\delta}{4}$ and $f_{n_0}(x_2) < r + \frac{\delta}{4}$ and from Lemma 10.3 it follows that f_{n_0} is not measurable—a contradiction.

(b) Category case. Let $Y \in \mathcal{B} \setminus \mathcal{K}$, $r \in \mathbb{R}$ and $\delta > 0$ be a set and numbers whose existence is guaranteed. Suppose that f has the Baire property. Put $Z_M = \{x \in Y\colon |f(x)| < M\}$. Obviously $\bigcup_{M=1}^{\infty} Z_M = Y$. There exists M_0 such that Z_{M_0} is of the second category (and has the Baire property). Let $\{f_n\}_{n\in\mathbb{N}}$ be a sequence of simple functions having the Baire property convergent $\mathcal{K}$-a.e. to f and convergent uniformly to f on Z_{M_0}. Let $n_0 \in \mathbb{N}$ be such that $\| f_{n_0} \| = \sup\{|f(x) - f_{n_0}(x)|\colon x \in Z_{M_0}\} < \frac{\delta}{4}$. If $A \subset Z_{M_0}$, $A \in \mathcal{B} \setminus \mathcal{K}$, then there exist $x_1, x_2 \in A$ such that $f(x_1) > r + \delta$, $f(x_2) < r$. Again a contradiction. □

As a immediate consequence of Theorem 10.4 we have:

Theorem 10.5 *If f is not $\mathcal{S}$-measurable function of accumulation of a sequence $\{f_n\}_{n\in\mathbb{N}}$, then there exist $Y \in \mathcal{S} \setminus \mathcal{I}$ and two numbers $r \in \mathbb{R}$ and $\delta > 0$ such that for each set $A \subset Y$, $A \in \mathcal{S} \setminus \mathcal{I}$ there exist two infinite sets $M_1, M_2 \subset \mathbb{N}$ and $x_1, x_2 \in A$ for which $f_n(x_1) > r + \delta$ for $n \in M_1$ and $f_n(x_2) < r$ for $n \in M_2$.*

Now we shall need the definition of Souslin operation (A) (see, for example [Sak37], p. 43). We call determining system any family $\mathfrak{A} = \{A_{n_1,n_2,\dots n_k}\}$ of sets such that with each finite sequence of positive integers $\{n_1, n_2, \dots n_k\}$ there is associated a set $A_{n_1,n_2,\dots n_k}$. The set

$$A = \bigcup_{n_1,n_2,\dots,n_k,\dots} A_{n_1} \cap A_{n_1 n_2} \cap \dots \cap A_{n_1 n_2,\dots,n_k} \cap \dots$$

where the summation extends over all infinite sequences $\{n_1, n_2, \dots, n_k, \dots\}$ of indices is called the nucleus of determining system $\mathfrak{A}$ and denoted $K(\mathfrak{A})$. The operation leading from a determining system to its nucleus is called the operation (A) or Souslin operation. Lusin and Sierpiński have proved in [LS18] that if a determining system $\mathfrak{A}$ consists of Lebesgue measurable sets, then $K(\mathfrak{A}) \in \mathcal{L}$. Nikodym [Nik25] has obtained similar result for sets having the Baire property. For more general theorem see [Szp33].

Theorem 10.6 *Suppose that $\{f_n\}_{n\in\mathbb{N}}$ is a sequence of $\mathcal{S}$-measurable real-valued functions defined on $[0, 1]$ and $M \subset \mathbb{N}$ is an infinite set of positive integers. For each $r \in \mathbb{R}$ and $\delta > 0$ the set $K_M(r, \delta)$ consisting of all points $x \in [0, 1]$ for which there exist two infinite disjoint subsets P_1 and P_2 of M such that $f_n(x) > r + \delta$ for $n \in P_1$ and $f_n(x) < r$ for $n \in P_2$ is $\mathcal{S}$-measurable.*

Proof For simplicity of denotation suppose that $M = \mathbb{N}$. For each finite sequence $\{n_1, n_2, \ldots, n_k\}$ of positive integers put

$$A_{n_1,n_2,\ldots,n_k} = \big\{x \in [0, 1]\colon\ f_{n_j}(x) > r + \delta \text{ for } j \in \{1, 2, \ldots, k\}\big\}.$$

The family $\mathfrak{A} = \{A_{n_1,n_2,\ldots,n_k}\}$ is a determining system consisting of $\mathcal{S}$-measurable sets, so $K_1 = K(\mathfrak{A})$ is $\mathcal{S}$-measurable. Similarly put

$$B_{n_1,n_2,\ldots,n_k} = \big\{x \in [0, 1]\colon\ f_{n_j}(x) < r \text{ for } j \in \{1, 2, \ldots, k\}\big\}$$

and $\mathfrak{B} = \{B_{n_1,n_2,\ldots,n_k}\}$. Then $K_2 = K(\mathfrak{B})$ is also $\mathcal{S}$-measurable. Obviously $K_M(r, \delta) = K_1 \cap K_2$. □

Recall that $M_1 \Subset M_2$ for $M_1, M_2 \subset \mathbb{N}$ means that $\operatorname{card}(M_1 \setminus M_2) < \aleph_0$.

Theorem 10.7 *If $\{f_n\}_{n\in\mathbb{N}}$ is a sequence of real-valued $\mathcal{S}$-measurable functions defined on $[0, 1]$ which has not a subsequence convergent $\mathcal{I}$-a.e., then there exist an infinite set $M_0 \subset \mathbb{N}$ and numbers $r \in \mathbb{R}$ and $\delta > 0$ such that for each $M \Subset M_0$ we have $K_M(r, \delta) \in \mathcal{S} \setminus \mathcal{I}$. Moreover, the set M_0 can be chosen in such a way that $K_{M_0}(r, \delta) \setminus K_M(r, \delta) \in \mathcal{I}$ for each $M \Subset M_0$.*

Proof Suppose that this is not the case. Let $\{r_n, \delta_n\}_{n\in\mathbb{N}}$ be a sequence consisting of all ordered pairs such that $r_n, \delta_n \in \mathbb{Q}$ and $\delta_n > 0$. By induction we can construct a sequence of infinite subsets of $\mathbb{N}$ such that $M_{n+1} \Subset M_n$ and $K_{M_n}(r_n, \delta_n) \in \mathcal{I}$ for each $n \in \mathbb{N}$. Let M_0 be an infinite subset of $\mathbb{N}$ such that $M_0 \Subset M_n$ for each n. We shall show that the sequence $\{f_n(x)\colon n \in M_0\}$ converges $\mathcal{I}$-a.e. Let $A = \{x \in [0, 1]\colon \{f_n\colon n \in M_0\} \text{ diverges}\}$. Suppose that $A \notin \mathcal{I}$. Then $A \setminus \bigcup_{n\in\mathbb{N}} K_{M_n}(r_n, \delta_n) \neq \emptyset$. If $x_0 \in A \setminus \bigcup_{n\in\mathbb{N}} K_{M_n}(r_n, \delta_n)$ then there exists a pair (r_{n_0}, δ_{n_0}) such that $\liminf_{n\in M_0} f_n(x_0) < r_{n_0} < r_{n_0} + \delta_{n_0} < \limsup_{n\in M_0} f_n(x_0)$. But $\limsup_{n\in M_0} f_n(x_0) \leq r_{n_0} + \delta_{n_0}$ or $\liminf_{n\in M_0} f_n(x_0) \geq r_{n_0}$, since $x_0 \notin K_{M_{n_0}}(r_{n_0}, \delta_{n_0})$, a contradiction.

If M_0 fulfills the second thesis, we are done. If not, there exists an infinite set $M_1 \Subset M_0$ for which $K_{M_0}(r, \delta) \setminus K_{M_1}(r, \delta) \notin \mathcal{I}$. Suppose that we have constructed sets M_β for $\beta < \alpha < \omega_1$ such that for each $\gamma < \beta < \alpha$ $M_\beta \Subset M_\gamma$ and $K_{M_\gamma}(r, \delta) \setminus K_{M_\beta}(r, \delta) \notin \mathcal{I}$. If $\alpha = \beta + 1$, there exists $M_\alpha \Subset M_\beta$ such that $K_{M_\beta}(r, \delta) \setminus K_{M_\alpha}(r, \delta) \notin \mathcal{I}$ (unless

M_β is the required set). If α is a limit number we construct M_α using the diagonal procedure and we obtain $M_\alpha \Subset M_\beta$ for each $\beta < \alpha$ and $K_{M_\alpha}(r, \delta) \subset \bigcap_{\beta<\alpha} K_{M_\beta}(r, \delta)$. Hence for each $\gamma \leq \alpha$ we have $K_{M_\alpha}(r, \delta) \setminus K_{M_\gamma}(r, \delta) \notin \mathcal{I}$. The procedure must stop at some $\alpha_0 < \omega_1$, since both $(\mathcal{L}, \mathcal{N})$ and $(\mathcal{B}, \mathcal{K})$ fulfill (ccc). So M_{α_0} fulfills all requirements. □

Theorem 10.8 ([BG78]) *If $\{f_n\}_{n\in\mathbb{N}}$ is a sequence of commonly bounded real-valued $\mathcal{S}$-measurable functions defined on* $[0, 1]$ *which has not a subsequence convergent $\mathcal{I}$-a.e., then there exists a function of accumulation of this sequence which is not $\mathcal{S}$-measurable.*

Proof For simplicity suppose that $|f_n(x)| \leq 1$ for each $x \in [0, 1]$ and $n \in \mathbb{N}$. Consider a subsequence $\{f_n\}_{n\in M_0}$, where M_0 is the set from the second part of Theorem 10.4. Recall that $K_{M_0}(r, \delta) \in \mathcal{S} \setminus \mathcal{I}$ for some $r \in \mathbb{R}$ and $\delta > 0$. Put $\mathcal{A} = \{A \subset K_{M_0}(r, \delta)\colon A \in \mathcal{L} \setminus \mathcal{N}$ and $A \in \mathcal{F}_\sigma\}$ (the measure case) and $\mathcal{A} = \{A \subset K_{M_0}(r, \delta)\colon A \in \mathcal{B} \setminus \mathcal{K}$ and $A \in \mathcal{G}_\sigma\}$ (the category case). In both cases $\mathrm{card}(\mathcal{A}) = \mathfrak{c}$ and for each $E \subset K_{M_0}(r, \delta)$, $E \in \mathcal{S} \setminus \mathcal{I}$ there exists $A \in \mathcal{A}$ such that $A \subset E$. Let $\{A_\alpha\colon \alpha < \omega_1\}$ be well-ordering of $\mathcal{A}$. We shall construct a transfinite sequence $\{M_\alpha\colon \alpha < \omega_1\}$ of subsets of M_0 such that $\mathrm{card}(M_\alpha \setminus M_\beta) < \aleph_0$ for $\beta < \alpha < \omega_1$ and two transfinite sequences $\{x_\alpha\colon \alpha < \omega_1\}$ and $\{y_\alpha\colon \alpha < \omega_1\}$ such that $x_\alpha, y_\alpha \in A_\alpha$ and $f_n(x_\alpha) > r + \delta$, $f_n(y_\alpha) < r$ for all $n \in M_\alpha$. Suppose that we have constructed M_β, x_β, y_β for $\beta < \alpha$. The family $\{M_\beta : \beta < \alpha\}$ is denumerable, so (in the case when α is a limit ordinal) using the diagonal procedure one can construct the set $M' \subset M_0$ such that $M' \Subset M_\beta$ for $\beta < \alpha$. According to Theorem 10.4 $K_{M_0}(r, \delta) \setminus K_{M'}(r, \delta) \in \mathcal{I}$, so $A_\alpha \cap K_{M'}(r, \delta) \in \mathcal{S} \setminus \mathcal{I}$. Hence there exist $x_\alpha \in A_\alpha \cap K_{M'}(r, \delta)$ and an infinite set $M'' \subset M'$ such that $f_n(x_\alpha) > r + \delta$ for each $n \in M''$. Since $A_\alpha \cap K_{M''}(r, \delta) \in \mathcal{S} \setminus \mathcal{I}$ there exist $y_\alpha \in A_\alpha$ and an infinite set $M_\alpha \subset M''$ such that $f_n(y_\alpha) < r$ for all $n \in M_\alpha$. This ends the construction of M_α, x_α and y_α if α is limit. If $\alpha = \beta + 1$ the construction is simpler. Let D_α be the set of all functions of accumulation of a sequence $\{f_n\}_{n\in M_\alpha}$ for $\alpha < \omega_1$ and D'_k—the set of all functions of accumulation of a sequence $\{f_n\}_{n\geq k}$ for $k \in \mathbb{N}$. For each $\alpha < \omega_1$ the set D_α is closed in $[-1, 1]^{[0,1]}$. Hence D_α is compact for each $\alpha < \omega_1$. Similarly $D'_k \subset [-1, 1]^{[0,1]}$ is compact for each $k \in \mathbb{N}$. The family $D_\alpha\colon \alpha < \omega_1\} \cup \{D'_k\colon k \in \mathbb{N}\}$ is scentered, so $\bigcap_{\alpha<\omega_1} D_\alpha \cap \bigcap_{k\in\mathbb{N}} D'_k \neq \emptyset$. If f belongs to this intersection, then it is a function of accumulation of $\{f_n\}_{n\in\mathbb{N}}$. From the construction it follows that f fulfills the condition from Theorem 10.1, so it is not $\mathcal{S}$-measurable. □

The above theorem shows only the existence of not $\mathcal{S}$-measurable function of accumulation, while in the Theorem of Fremlin all functions of accumulation of some sequences are not measurable. However, this weaker result is sufficiently strong to reformulate the next theorem in the following way:

Theorem 10.9 *If $\{f_n\}_{n\in\mathbb{N}}$ is a sequence is a sequence of commonly bounded real-valued $\mathcal{S}$-measurable functions defined on* $[0, 1]$ *such that each function of accumulation of this sequence is $\mathcal{S}$-measurable, then there exists a subsequence $\{f_{n_m}\}_{m\in\mathbb{N}}$ of $\{f_n\}_{n\in\mathbb{N}}$ which converges $\mathcal{I}$-a.e.*

Remark 10.10 The theorem of Fremlin is much more general, because it works in a perfect measure space. Recall that measure space $(X, \mathcal{S}, \mu)$ is perfect if for each measurable function $f : X \to \mathbb{R}$, each set $E \in \mathcal{S}$ and each set $A \subset \mathbb{R}$ such that $f^{-1}(A) \in \mathcal{S}$ there exists a Borel set $B \subset A$ such that $\mu(E \cap f^{-1}(A)) = \mu(E \cap f^{-1}(B))$. The notion od perfect measure space has been introduced by Sazonov [Saz65]. Talagrand [Ta-d80] has constructed a measure space $(X, \mathcal{S}, \mu)$ and a sequence of $\mathcal{S}$-measurable real-valued functions which is relatively compact but does not contain any sequence convergent μ-almost everywhere.

The paper of Bourgain et al. [BFT78] is devoted to the study of sets of real-valued functions defined on a topological space, measurable with respect to some σ-algebra of sets and satisfying some criterion of compactness or relative compactness in the topology of pointwise convergence. The theorem concerning functions with the Baire property, again more general than Theorem 10.9, looks as follows: Let X be a Polish space, $B(X)$—the space of real-valued functions on X with the Baire property. If A is a countable relatively countably compact set in $B(X)$ and for each $f \in \overline{A}$ there is a sequence in A convergent of f except on a set of the first category. The terms concerning compactness in the above theorems are in accordance with the terminology [Eng89].

Chapter 11
Sequences of Functions of Two Variables

In this chapter we shall show that the behaviour of a sequence $\{f_n\}_{n\in\mathbb{N}}$ of functions defined on the unit square such that $\{f_n(\cdot, y)\}_{n\in\mathbb{N}}$ converges a.e. to f for a.e $y \in [0, 1]$ is the same as in the case of a sequence $\{f_n\}_{n\in\mathbb{N}}$ of functions having the Baire property such that $\{f_n(\cdot, y)\}_{n\in\mathbb{N}}$ converges $\mathcal{K}$-a.e. for $\mathcal{K}$-a.e. $y \in [0, 1]$, but for the convergence in measure and in category situation is completely different.

Recall that if $(X, \mathcal{S})$ is a measurable space and $\{f_n\}_{n\in\mathbb{N}}$ is a sequence of real-valued $\mathcal{S}$-measurable functions defined on X, then the set $A = \{x \in X: \{f_n(x)\}_{n\in\mathbb{N}} \text{ converges}\}$ belongs to $\mathcal{S}$. Indeed,

$$A = \bigcap_{k\in\mathbb{N}} \bigcup_{n\in\mathbb{N}} \bigcap_{i,j\geq n} \left\{x \in A: |f_i(x) - f_j(x)| < \frac{1}{k}\right\}.$$

Also, if $f: A \to \mathbb{R}$ is defined by $f(x) = \lim_{n\to\infty} f_n(x)$ for $x \in A$, then f is $\mathcal{S}$-measurable, because

$$\{x \in A: f(x) < a\} = \bigcap_{k\in\mathbb{N}} \bigcup_{n\in\mathbb{N}} \bigcap_{i\geq n} \left\{x \in A: f_n(x) \leq a - \frac{1}{k}\right\}.$$

Suppose now that all real-valued functions under consideration are defined in the unit square $[0, 1] \times [0, 1]$ equipped with a σ-algebra $\mathcal{L}_2$ of Lebesgue measurable sets and two-dimensional Lebesgue measure λ_2.

Theorem 11.1 *If $\{f_n\}_{n\in\mathbb{N}}$ is a sequence of real-valued functions defined on $[0, 1] \times [0, 1]$, $\mathcal{L}_2$-measurable such that the sequence $\{f_n(\cdot, y)\}_{n\in\mathbb{N}}$ converges λ-a.e. to some function f for λ-almost every $y \in [0, 1]$, then $\{f_n\}_{n\in\mathbb{N}}$ converges to f λ_2-a.e. and f is $\mathcal{L}_2$-measurable.*

W. Wilczyński, *Convergence in Measure and in Category*, Lecture Notes in Mathematics 2394, https://doi.org/10.1007/978-3-032-23143-7_11

Proof If

$$A = \{(x, y) \in [0, 1] \times [0, 1] \colon \{f_n(x, y)\}_{n\in\mathbb{N}} \text{ converges to } f(x, y)\}$$

and $A^y = \{x \in [0, 1] \colon (x, y) \in A\}$, then $\lambda(A^y) = 1$ for a.e. $y \in [0, 1]$, so $\lambda_2(A) = 1$ by virtue of Fubini theorem. □

Theorem 11.2 *If $\{f_n\}_{n\in\mathbb{N}}$ is a sequence of real-valued functions defined on $[0, 1] \times [0, 1]$ having the Baire property ($\mathcal{B}_2$-measurable) such that the sequence $\{f_n(\cdot, y)\}_{n\in\mathbb{N}}$ converges $\mathcal{K}$-a.e. to some function f for $\mathcal{K}$-almost every $y \in [0, 1]$, then $\{f_n\}_{n\in\mathbb{N}}$ converges to f $\mathcal{K}_2$-a.e.*

Proof If

$$A = \{(x, y) \in [0, 1] \times [0, 1] \colon \{f_n(x, y)\}_{n\in\mathbb{N}} \text{ converges to } f(x, y)\}$$

then A^y is residual for each $y \in [0, 1]$ except on a set of first category. Then A is residual in $[0, 1] \times [0, 1]$ by virtue of Kuratowski-Ulam theorem. □

When we are dealing with the convergence everywhere except on a set from some σ-ideal (sets of measure zero, sets of first category), the behaviour of measurable functions and functions having the Baire property is similar. Below we shall show that the situation is different for convergence in measure and in category.

Theorem 11.3 ([Orl51]) *If $\{f_n\}_{n\in\mathbb{N}}$ is a sequence of real-valued, $\mathcal{L}_2$-measurable functions defined on $[0, 1] \times [0, 1]$ such that the sequence $\{f_n(\cdot, y)\}_{n\in\mathbb{N}}$ converges in measure to some function f for λ-almost every $y \in [0, 1]$, then $\{f_n\}_{n\in\mathbb{N}}$ converges in measure λ_2 to f.*

Proof Put

$$h_n(y) = \int_0^1 \frac{|f_n(x, y) - f(x, y)|}{1 + |f_n(x, y) - f(x, y)|} dx.$$

from the assumption it follows that $\lim_{n\to\infty} h_n(y) = 0$ for λ-a.e. $y \in [0, 1]$ (see Theorem 2.11). Hence $\lim_{n\to\infty} \int_0^1 h_n(y)dy = 0$. Since

$$\int_0^1 h_n(y)dy = \int_0^1 \int_0^1 \frac{|f_n(x, y) - f(x, y)|}{1 + |f_n(x, y) - f(x, y)|} dxdy,$$

again from Theorem 2.11 it follows that $\{f_n\}_{n\in\mathbb{N}}$ converges in measure λ_2 to f. □

Theorem 11.4 ([Wil20]) *There exists a sequence of real functions having the Baire property $\{f_n\}_{n\in N}$ defined on the unit square such that for each $y \in [0, 1]$ the sequence $\{f_n(\cdot, y)\}_{n\in N}$ converges in category to 0 but $\{f_n\}_{n\in N}$ does not converge in category.*

Proof For $n \in N$ let

$$I_i^n = \left(\frac{i}{n}, \frac{i+1}{n}\right) \quad \text{for } i \in \{0, 1, \ldots, n-1\}$$

$$J_i^n = \left(\frac{i}{n^2}, \frac{i+1}{n^2}\right) \quad \text{for } i \in \{0, 1, \ldots, n^2-1\}.$$

Put $A_n = \bigcup_{i=0}^{n-1} \left(I_i^n \times \bigcup_{j=0}^{n-1} J_{j \cdot n+i}^n\right)$ and $f_n = \chi_{A_n}$ for $n \in N$.

If $y \in [0, 1]$, then $f_n(\cdot, y)$ is either equal to 0 or is a characteristic function of the interval of length $\frac{1}{n}$, so for each $y \in [0, 1]$ the sequence $\{f_n(\cdot, y)\}_{n \in N}$ converges to 0 in category, because from each subsequence one can choose the subsequence convergent to 0 everywhere or except on one-point set.

Observe now that for each increasing sequence $\{n_m\}_{m \in N}$ of positive integers and for each $p \in N$ both sets $\bigcup_{m=p}^{\infty} A_{n_m}$ and $\bigcup_{m=p}^{\infty}(([0, 1] \times [0, 1]) \setminus A_{n_m})$ include open dense sets, from which it easily follows that $\{f_n\}_{n \in N}$ does not converge in category to any function. □

Observe also that if $x \in [0, 1]$ is irrational, then for $n \in N$ the function $f_n(x, \cdot)$ is a characteristic function of the union of n open intervals equidistributed on $[0, 1]$. If we denote for such x $B_n^x = \{y\colon f_n(x, y) = 1\}$, then for each increasing sequence $\{n_m\}_{m \in N}$ and for each $p \in N$ we see that $\bigcup_{m=p}^{\infty} B_{n_m}^x$ and $\bigcup_{m=p}^{\infty}[0, 1] \setminus B_{n_m}^x$ include open dense sets, so $\{f_n(x, \cdot)\}_{n \in N}$ does not converge in category to any function for irrational $x \in [0, 1]$. The next theorem will show that even if the sequence $\{f_n\}_{n \in N}$ converges in category to 0 on each segment connecting points of the boundary of $[0, 1] \times [0, 1]$ (with respect to the topology of the segment), then it can happen that $\{f_n\}_{n \in N}$ does not converge in category.

Theorem 11.5 ([Wil20]) *There exists a sequence $\{g_n\}_{n \in N}$ of real functions having the Baire property defined on the unit square such that for each segment D connecting points of the boundary of $[0, 1] \times [0, 1]$ the sequence $\{g_n|D\}_{n \in N}$ (treated as the sequence of functions of one variable) consists of functions having the Baire property on D and converges in category to 0 but $\{g_n\}_{n \in N}$ does not converges in category.*

Proof We shall need the following lemma, in which $K(x, r) = \{p\colon d(p, x) \leq r\}$.

Lemma 11.6 *If $E \subset [0, 1] \times [0, 1]$ is a finite set such that no three points of E are collinear, then there exists $k \in N$ such that each segment connecting points of the boundary of $[0, 1] \times [0, 1]$ has points in common with at most two circles from the family $\mathcal{E}_k = \{\{K(x, k^{-1})\}\colon x \in E\}$.*

Proof of the Lemma Suppose that this not the case, so for each $k \in N$ there exists a segment D_k connecting points of the boundary of $[0, 1] \times [0, 1]$ and intersecting at least three discs belonging to $\mathcal{E}_k$. Then there exist three different points $p_1, p_2, p_3 \in E$ and an increasing sequence $\{k_m\}_{m \in N}$ of positive integers such that $D_{k_m} \cap K(\pi_i, k_m^{-1}) \neq 0$ for $i = 1, 2, 3$ and for each $m \in N$. Observe that $D_{k_m} \xrightarrow[m\to\infty]{} D_0$ in Hausdorff metric in $\mathbb{R}^2$ and then $p_1, p_2, p_3 \in D_0$, (are collinear)—a contradiction. □

Let's return to the proof of theorem. Let $E_n = \{p_0, p_1, \ldots, p_{n^2-1}\} \subset [0, 1] \times [0, 1]$ be a set such that card $E_n \cap \left(\left(\frac{i}{n}, \frac{i+1}{n}\right) \times \left(\frac{j}{n}, \frac{j+1}{n}\right)\right) = 1$ for each $i, j \in 0, 1, \ldots, n-1$ and no three points of E_n are collinear.

Let m_n (for each $n \in N$) be a number described in the Lemma. Obviously $m_n \xrightarrow[n\to\infty]{} \infty$. Put $A_n = \bigcup_{i=0}^{n^2-1} K(p_i, m_n^{-1})$ and $g_n = \chi_{A_n}$.

If D is an arbitrary segment connecting points of the boundary of the unit square, then $g_n|D$ in equal to 0 or is a characteristic function of the set consisting of one or two intervals, each of the length less than $2 \cdot m_m^{-1}$, so $\{g_n|_D\}_{n\in N}$ converges in category to 0.

The proof that $\{g_n\}_{n\in N}$ does not converge in category is similar to that for $\{f_n\}_{n\in N}$ from Theorem 11.4. □

Chapter 12
Essential Upper Bound of a Measurable Periodic Function

Steinhaus [Sth20] has proved that if $\{a_n\}_{n\in\mathbb{N}}$ and $\{b_n\}_{n\in\mathbb{N}}$ are arbitrary sequences of real numbers, then

$$\limsup_{n\to\infty} |a_n \cos nx + b_n \sin nx| = \limsup_{n\to\infty} \sqrt{a_n^2 + b_n^2}$$

almost everywhere. Mazur and Orlicz have found in [MO40] a property of periodic measurable functions from which the result of Steinhaus follows immediately. Recall that if $f\colon \mathbb{R} \to \mathbb{R}$ is measurable function, then $\sup\text{ess}\, f = \inf\{y \in \mathbb{R}\colon \lambda(\{x \in \mathbb{R}\colon f(x) > y\}) = 0\}$ if there exists $y \in \mathbb{R}$ such that $\lambda(\{x \in \mathbb{R}\colon f(x) > y\}) = 0$. If $\lambda(\{x \in \mathbb{R}\colon f(x) > y\}) > 0$ for each $y \in \mathbb{R}$, then we put $\sup\text{ess}\, f = +\infty$.

In the sequel we shall suppose that $\{w_n\}_{n\in\mathbb{N}}$ is a sequence of real numbers such that $\lim_{n\to\infty} w_n = +\infty$ and $\{t_n\}_{n\in\mathbb{N}}$ is an arbitrary sequence of real numbers.

Theorem 12.1 ([MO40, Orl48]) *If $f\colon \mathbb{R} \to \mathbb{R}$ is a measurable periodic function, then $\limsup_{n\to\infty} f(w_n x + t_n) = \sup\text{ess}\, f$ almost everywhere.*

Proof It is sufficient to prove the theorem for periodic functions with period 1. Suppose first that $\sup\text{ess}\, f < +\infty$. Let

$$E_k = \left\{x \in \mathbb{R}\colon f(x) > \sup\text{ess}\, f - \frac{1}{2k}\right\} \text{ for } k \in \mathbb{N}$$

and

$$E_k^n = \left\{x \in \mathbb{R}\colon \sup(f(w_n x + t_n), f(w_{n+1} x + t_{n+1}), \ldots) \le \sup\text{ess}\, f - \frac{1}{k}\right\}$$

for $k, n \in \mathbb{N}$. Obviously E_n^k, E_k are measurable, $\lambda(E_k \cap (0,1)) > 0$ and $E_k^n \cap E_k = \emptyset$ for $k, n \in \mathbb{N}$. We shall show that $\lambda(E_k^n) = 0$ for each $k, n \in \mathbb{N}$. Suppose that this is

W. Wilczyński, *Convergence in Measure and in Category*, Lecture Notes in Mathematics 2394, https://doi.org/10.1007/978-3-032-23143-7_12

not the case, so there exist $k_0, n_0 \in \mathbb{N}$ such that $\lambda(E_{k_0}^{n_0}) > 0$. Let x_0 be a point of density of $E_{k_0}^{n_0}$. For each $i \in \mathbb{N}$ there exists $l(i) \in \mathbb{Z}$ such that $\frac{l(i)-1}{w_i} \le x_0 \le \frac{l(i)}{w_i}$. Put $I_{l(i)} = \left(\frac{l(i)-1}{w_i}, \frac{l(i)+1}{w_i}\right)$. If $v \in \left(1 - \frac{1}{2}\lambda(E_{k_0} \cap (0, 1)), 1\right)$, then there exists $i \ge n$ for which $\lambda\left(E_{k_0}^{n_0} \cap I_{l(i)}\right) \ge v\lambda\left(I_{l(i)}\right)$. Let $g_{n,l}(x) = \frac{x-t_n}{w_n} + \frac{l}{w_n}$ for $n \in \mathbb{N}, l \in \mathbb{Z}$. We have $g_{n,l}\left(E_{k_0} \cap (0, 1)\right) \subset \left(\frac{l-[t_n]-1}{w_i}, \frac{l-[t_n]+1}{w_i}\right)$, so if we put $p(i) = l(i) + [t_i]$, then $g_{i,p(i)}\left(E_{k_0} \cap (0, 1)\right) \subset I_{l(i)}$. Since

$$\lambda\left(g_{i,p(i)}\left(E_{k_0} \cap (0, 1) \cap I_{l(i)}\right)\right) = \frac{1}{w_i}\lambda\left(E_{k_0} \cap (0, 1)\right) = \frac{1}{2}\lambda\left(I_{l(i)}\right) \cdot \lambda\left(E_{k_0} \cap (0, 1)\right),$$

we have

$$\lambda\left(E_{k_0}^{n_0} \cap I_{l(i)}\right) + \lambda\left(g_{i,p(i)}\left(E_{k_0} \cap (0, 1) \cap I_{l(i)}\right)\right) >$$
$$> \frac{1}{2}\lambda\left(I_{l(i)}\right) \cdot \lambda\left(E_{k_0} \cap (0, 1)\right) + v\lambda\left(I_{l(i)}\right) > \lambda\left(I_{l(i)}\right).$$

Hence $E_{k_0}^{n_0} \cap g_{i,p(i)}\left(E_{k_0} \cap (0, 1)\right) \ne \emptyset$. So there exists $x_1 \in E_{k_0}^{n_0}$ such that $x_1 \in g_{i,p(i)}\left(E_{k_0} \cap (0, 1)\right)$, which means that for some $z \in E_{k_0} \cap (0, 1)$ we have $x_1 = g_{i,p(i)}(z)$. Hence $z = w_i x_1 + t_i - p(i)$ and $f(w_i x_1 + t_i) = f\left(w_i x_1 + t_i - p(i)\right) > \text{sup ess } f - \frac{1}{2k_0}$, which is a contradiction with $x_1 \in E_{k_0}^{n_0}$. This proves that $\lambda\left(E_k^n\right) = 0$ for each $k, n \in \mathbb{N}$. Put $H = \bigcup_{k,n\in\mathbb{N}} E_k^n$. Obviously $\lambda(H) = 0$. If $x \notin H$, then $\limsup_{n\to\infty} f(w_n x + t_n) \ge \text{sup ess } f$. From the definition of sup ess f it follows that there exists a set C such that $\lambda(C) = 0$ and for each $x \in C$ $\sup(f(w_n x + t_n), f(w_{n+1}x + t_{n+1}), \ldots) \le \text{sup ess } f$ which ends the proof when $\text{sup ess } f < +\infty$. If $\text{sup ess } f = +\infty$ it suffices to consider $F = \text{arctg } f$. □

Corollary 12.2 ([MO40]) *If $f\colon \mathbb{R} \to \mathbb{R}$ is a measurable periodic function and $\{a_n\}_{n\in\mathbb{N}}$—an arbitrary sequence of real numbers, then*

$$\limsup_{n\to\infty} a_n f(w_n x + t_n) = \text{sup ess } f \cdot \limsup_{n\to\infty} |a_n|$$

almost everywhere.

Proof Let $\{n_i\}_{i\in\mathbb{N}}$ be an increasing sequence of positive integers such that $\limsup_{n\to\infty} |a_n| = \lim_{i\to\infty} |a_{n_i}|$. Then

$$\limsup_{n\to\infty} |a_n f(w_n x + t_n)| \ge \limsup_{i\to\infty} |a_{n_i} f(w_{n_i} x + t_{n_i})| =$$
$$= \limsup_{i\to\infty} |f(w_{n_i} x + t_{n_i})| \cdot \lim_{i\to\infty} |a_{n_i}| = \text{sup ess } f \cdot \limsup_{n\to\infty} |a_n|$$

(the last equality holds almost everywhere). The converse inequality is obvious. □

Corollary 12.3 ([MO40]) *If $E \subset [0,1)$ is a measurable set such that $\lambda(E) > 0$, then for almost each x there exists an increasing sequence $\{n_i\}_{i\in\mathbb{N}}$ of positive integers such that $w_{n_i}x + t_{n_i} - [w_{n_i}x + t_{n_i}] \in E$ for each $i \in \mathbb{N}$.*

Proof Put $E_0 = \bigcup_{k\in\mathbb{Z}}(E+k)$ and $f = \chi_{E_0}$. A function f is measurable and periodic with period 1, so $\limsup_{n\to\infty} f(w_n x + t_n) = 1$ almost everywhere, which ends the proof. □

Remark 12.4 The above result was proved by Hardy and Littlewood [HL14] in another way.

The above theorem and corollaries have category analogues. It can be proved using similar ideas. It is worth observing that in the paper [MO40] the essential role was played by the Lebesgue integral.

If $f : \mathbb{R} \to \mathbb{R}$ is a function having the Baire property, then

$$\text{sup cat } f = \inf\{y \in \mathbb{R}\colon \{x \in \mathbb{R}\colon f(x) > y\} \text{ is of the first category}\}$$

if there exists $y \in \mathbb{R}$ such that $\{x \in \mathbb{R}\colon f(x) > y\}$ is of the first category. If $\{x \in \mathbb{R}\colon f(x) > y\}$ is of the second category for each $y \in \mathbb{R}$, we put sup cat $f = +\infty$. The sequences $\{w_n\}_{n\in\mathbb{N}}$ and $\{t_n\}_{n\in\mathbb{N}}$ are as before.

Theorem 12.5 ([Orl48]) *If $f : \mathbb{R} \to \mathbb{R}$ is a function having the Baire property and periodic, then* $\limsup_{n\to\infty} f(w_n x + t_n) =$ sup cat f *$\mathcal{K}$-almost everywhere.*

Proof We shall suppose that f has period 1 and sup cat $f < 1$. Let E_k^n, E_k and $g_{n,l}$ be defined as in the previous proof. Hence E_k^n for each $k, n \in \mathbb{N}$ and E_k have the Baire property and E_k is of the second category for each $k \in \mathbb{N}$. We shall show that E_k^n is of the first category for each $k, n \in \mathbb{N}$. Suppose that it is not the case, so there exist $k_0, n_0 \in \mathbb{N}$ such that $E_{k_0}^{n_0}$ is of a second category. There exists an interval $I_1 \subset \mathbb{R}$ such that $I_1 \setminus E_{k_0}^{n_0} \in \mathcal{I}$. Since E_{k_0} is also of the second category, there exists an interval $I_2 \subset \mathbb{R}$ such that $I_2 \setminus E_{k_0} \in \mathcal{I}$. If $i \geq n$, then there exists $l(i) \in \mathbb{N}$ such that $g_{i,l(i)}(I_2) \subset I_1$.

The function g preserves the Baire category, so $g_{i,l(i)}(I_2 \setminus E_{k_0}) \in \mathcal{I}$. Hence $g_{i,l(i)}(I_2 \setminus E_{k_0}) \cup (I_1 \setminus E_{k_0}^{n_0}) \in \mathcal{I}$, so $E_{k_0}^{n_0} \cap g_{i,l(i)}(I_2 \cap E_{k_0}^{n_0}) \neq \emptyset$. If x_0 belongs to this set, then $f(w_i x_0 + t_i) \leq$ sup cat $f - \frac{1}{k}$, because $i \geq n$ and $f(w_i x_0 + t_i) >$ sup cat $f - \frac{1}{2k}$, a contradiction. So each $E_k^n \in \mathcal{I}$ and $H = \bigcup_{k,n\in\mathbb{N}} E_k^n \in \mathcal{I}$. The rest of the proof is the same as in the measure case. □

Corollary 12.6 *If $E \subset (0,1)$ is a set of the second category having the Baire property, then for $\mathcal{K}$-almost each x there exists an increasing sequence $\{n_i(\cdot, y)\}_{i\in\mathbb{N}}$ of positive integers such that $w_{n_i}x + t_{n_i} - [w_{n_i}x + t_{n_i}] \in E$ for each $i \in \mathbb{N}$.*

Corollary 12.7 *If $f : \mathbb{R} \to \mathbb{R}$ is a periodic function having the Baire property and and $\{a_n\}_{n\in\mathbb{N}}$—an arbitrary sequence of real numbers, then*

$$\limsup_{n\to\infty} a_n f(w_n x + t_n) = \text{sup cat } f \cdot \limsup_{n\to\infty} |a_n|$$

$\mathcal{K}$-almost everywhere.

Proofs of both corollaries are almost the same as that of Corollaries 12.2 and 12.3. For almost periodic functions we have the following pair of theorems.

Theorem 12.8 *If f_1, f_2 are real-valued measurable periodic functions of real variable with periods p_1, p_2 such that $\frac{p_1}{p_2}$ is irrational, then* $\sup\text{ess}(f_1 + f_2) = \sup\text{ess}\, f_1 + \sup\text{ess}\, f_2$.

Proof Let

$$A_k = \left\{x \in \mathbb{R}\colon f_1(x) > \sup\text{ess}\, f_1 - \frac{1}{2k}\right\}$$

$$B_k = \left\{x \in \mathbb{R}\colon f_2(x) > \sup\text{ess}\, f_2 - \frac{1}{2k}\right\} \text{ for } k \in \mathbb{N}$$

Obviously $\lambda(A_k) > 0$ and $\lambda(B_k) > 0$. If x_1 is a point of density of A_k and x_2 is a point of density of B_k, then there exists $\delta > 0$ such that for each $h \in (0, \delta)$ we have

$$\lambda\left(A_k \cap (x_1, x_1 + h)\right) > \frac{3}{4}k, \qquad \lambda\left(A_k \cap (x_1 - h, x_1)\right) > \frac{3}{4}k,$$

$$\lambda\left(B_k \cap (x_2, x_2 + h)\right) > \frac{3}{4}k, \qquad \lambda\left(B_k \cap (x_2 - h, x_2)\right) > \frac{3}{4}k.$$

Observe that if $x_1 \neq x_2$, then either $\frac{x_2 - x_1}{p_1}$ or $\frac{x_2 - x_1}{p_2}$ is not an integer. Indeed, suppose that $\frac{x_2 - x_1}{p_1} = i$ and $\frac{x_2 - x_1}{p_2} = j$, where $i, j \in \mathbb{Z}$. Then $\frac{p_2}{p_1} = \frac{i}{j}$ – a contradiction. Suppose that $\frac{x_2 - x_1}{p_2}$ is not an integer. We can take sufficiently small $\delta > 0$ for which there exists $l \in \mathbb{Z}$ such that $\left(\frac{x_2 - x_1 - \delta}{p_2}, \frac{x_2 - x_1 + \delta}{p_2}\right) \subset (l, l+1)$. To avoid misunderstanding let $\varphi(t)$ denotes the fractional part of $t \in \mathbb{R}$ (instead of usual (t)). Then $\varphi\left(\frac{x_2 - x_1 - \delta}{p_2}, \frac{x_2 - x_1 + \delta}{p_2}\right) = \left(\varphi\left(\frac{x_2 - x_1 - \delta}{p_2}\right), \varphi\left(\frac{x_2 - x_1 + \delta}{p_2}\right)\right)$. From the classical theorem of Kronecker it (see for example [HWr62]) follows that there exist $m, n \in \mathbb{Z}$ such that $\frac{mp_1}{p_2} - n \in \left(\varphi\left(\frac{x_2 - x_1 - \delta}{p_2}\right), \varphi\left(\frac{x_2 - x_1 + \delta}{p_2}\right)\right)$. Then for $h = x_1 + mp_1 - x_2 - np_2$ we have $h \in (0, \delta)$, so

$$\lambda\left(B_k \cap (x_2 + np_2, x_2 + np_2 + h)\right) = \lambda\left(B_k \cap (x_2 + np_2, x_1 + mp_1)\right) > \frac{3}{4}h$$

and

$$\lambda\left(A_k \cap (x_1 + mp_1 - h, x_1 + mp_1)\right) = \lambda\left(A_k \cap (x_2 + np_2, x_1 + mp_1)\right) > \frac{3}{4}h.$$

Hence $\lambda(A_k \cap B_k) > 0$, so $\sup\text{ess}(f_1 + f_2) \geq \sup\text{ess}\, f_1 + \sup\text{ess}\, f_2 - \frac{1}{k}$. From the arbitrariness of $k \in \mathbb{N}$ it follows that $\lambda(A_k \cap B_k) > 0$, so $\sup\text{ess}(f_1 + f_2) \geq \sup\text{ess}\, f_1 + \sup\text{ess}\, f_2$. If $x_1 = x_2$ the proof is similar.

Now observe that $f_1 \le \sup\text{ess } f_1$ and $f_2 \le \sup\text{ess } f_2$ for almost each $x \in \mathbb{R}$, so $f_1 + f_2 \le \sup\text{ess } f_1 + \sup\text{ess } f_2$ a.e. and finally $\sup\text{ess}(f_1 + f_2) \le \sup\text{ess } f_1 + \sup\text{ess } f_2$. □

In a similar way one can prove

Theorem 12.9 *If f_1, f_2 are real-valued measurable periodic functions of real variable having the Baire property with periods p_1, p_2 such that $\frac{p_1}{p_2}$ is irrational, then* $\sup\text{cat}(f_1 + f_2) = \sup\text{cat } f_1 + \sup\text{cat } f_2$.

Chapter 13
Double Sequences of Measurable Functions

Sierpiński [Sie50] has proved that if $\{f_{m,n}\}_{m,n\in\mathbb{N}}$ is a double sequence of real-valued measurable functions defined on $[0, 1]$ such that for each sequences $\{m_k\}_{k\in\mathbb{N}}$ and $\{n_k\}_{k\in\mathbb{N}}$ of positive integers convergent to infinity the sequence $\{f_{m_k,n_k}\}_{k\in\mathbb{N}}$ converges a.e., then the double sequence $\{f_{m,n}\}_{m,n\in\mathbb{N}}$ converges a.e.. In the chapter we shall show general theorem due to Wagner [Wag81] concerning double sequences of functions.

Suppose that X is an arbitrary non-empty set, $\mathcal{S}$—a σ-algebra of subsets of X and $\mathcal{I} \subset \mathcal{S}$—a σ-ideal.

Definition 13.1 We shall say that the pair $(\mathcal{S}, \mathcal{I})$ fulfills the condition (F) if and only if for each $D \subset \mathcal{S} \setminus \mathcal{I}$ and for each double sequence $\{B_{j,n}\}_{j,n\in\mathbb{N}}$ of sets from $\mathcal{S}$ such that

(a) $B_{j,n} \subset B_{j,n+1}$ for each $j, n \in \mathbb{N}$
(b) $\bigcup_{n=1}^{\infty} B_{j,n} = D$ for each $j \in \mathbb{N}$
(c) $B_{j_1,n_1} \supset B_{j_2,n_2}$ if $j_1 < j_2$ and $j_1 + n_1 = j_2 + n_2$

there exists a sequence $\{n_j\}_{j\in\mathbb{N}}$ of positive integers such that $\limsup_{j\to\infty} B_{j,n_j} \notin \mathcal{I}$.

Remark 13.2 If the pair $(\mathcal{S}, \mathcal{I})$ fulfills the condition (E) (see Definition 1.17), then it fulfills also the condition (F).

Remark 13.3 If the double sequence of $\mathcal{S}$-measurable sets $\{B_{j,n}\}_{j,n\in\mathbb{N}}$ fulfills (a) and (b) with respect to a set D, then the double sequence $\{A_{j,n}\}_{j,n\in\mathbb{N}}$ defined by the formula $A_{j,n} = \bigcap_{i=1}^{j-1} B_{j-1,n+i}$ consists of $\mathcal{S}$-measurable sets and fulfills (a), (b) and (c) with respect to D.

Theorem 13.4 ([Wag81]) *Let $\{f_{m,n}\}_{m,n\in\mathbb{N}}$ be a double sequence of real-valued $\mathcal{S}$-measurable functions defined on X. The convergence $\mathcal{I}$-a.e. to f of all sequences $\{f_{m_k,n_k}\}_{k\in\mathbb{N}}$, where $\lim_{k\to\infty} m_k = \lim_{k\to\infty} n_k = +\infty$ implies the convergence $\mathcal{I}$-a.e. to f of a double sequence $\{f_{m,n}\}_{m,n\in\mathbb{N}}$ if and only if the pair $(\mathcal{S}, \mathcal{I})$ fulfills the condition (F).*

W. Wilczyński, *Convergence in Measure and in Category*, Lecture Notes in Mathematics 2394, https://doi.org/10.1007/978-3-032-23143-7_13

Proof Sufficiency. We can assume that $f(x) = 0$ for $x \in X$. Suppose that

$$E = \big\{x \in X : \{f_{m,n}\}_{m,n\in\mathbb{N}} \text{ does not converge to } 0\big\} \notin \mathcal{I}.$$

We shall show that there exist sequences $\{m_k\}_{k\in\mathbb{N}}$ and $\{n_k\}_{k\in\mathbb{N}}$ of positive integers convergent to infinity such that

$$\big\{x \in X : \{f_{m_k,n_k}\}_{k\in\mathbb{N}} \text{ does not converge to } 0\big\} \notin \mathcal{I}.$$

For each $x \in E$ there exist a number $a_x > 0$ and two sequences $\{m_k(x)\}_{k\in\mathbb{N}}$ and $\{n_k(x)\}_{k\in\mathbb{N}}$ such that $|f_{m_k(x),n_k(x)}| \geq a_x$ for each $k \in \mathbb{N}$. Put $E_a = \{x \in X : a_x \geq a\}$. Obviously $E = \bigcup_{i=1}^{\infty} E_{\frac{1}{i}}$. There exists $i_0 \in \mathbb{N}$ such that $E_{\frac{1}{i_0}} \notin \mathcal{I}$. Denote $E_{\frac{1}{i_0}} = D$ and put $H_{m,p} = \Big\{x \in X : |f_{m,p}(x)| \geq \frac{1}{i_0}\Big\} \cap D$ for $m, p \in \mathbb{N}$. If $B_{j,n} = \bigcup_{m=j}^{j+n-1} \bigcup_{p=j}^{j+n-1} H_{m,p}$ for $j, n \in \mathbb{N}$, then the double sequence $\{B_{j,n}\}_{j,n\in\mathbb{N}}$ fulfills (a), (b) and (c) with respect to D. From the condition (F) it follows that there exists a sequence $\{n_j\}_{j\in\mathbb{N}}$ of positive integers such that $B = \limsup_{j\to\infty} B_{j,n_j} \notin \mathcal{I}$. Let (m_k, n_k) be the kth element of the following sequence: $(1, 1)$, ..., $(1, 1 + n_1 - 1)$, $(2, 1)$, ..., $(2, 1 + n_1 - 1)$, ..., $(1 + n_1 - 1, 1)$, ..., $(1 + n_1 - 1, 1 + n_1 - 1)$, $(2, 2)$, ..., $(2, 2 + n_1 - 1)$, $(3, 2)$, ..., $(3, 2 + n_1 - 1)$, ..., $(2 + n_2 - 1, 2)$, $2, 2$, ..., $(2 + n_2 - 1, 2 + n_2 - 1)$, ..., (j, j), ..., $(j, j + n_j - 1)$, $(j + 1, j)$, ..., $(j + 1, j + n_j - 1)$, ..., $(j + n_j - 1, j)$, ..., $(j + n_j - 1, j + n_j - 1)$, ... Then $\lim_{k\to\infty} m_k = \lim_{k\to\infty} n_k = +\infty$. If $x \in B$, then in the sequence $\{|f_{m_k,n_k}|\}_{k\in\mathbb{N}}$ there are infinitely many terms not less than $\frac{1}{i_0}$, so the sequence $\{f_{m_k,n_k}\}_{k\in\mathbb{N}}$ does not converge to zero $\mathcal{I}$-a.e. on X.

Necessity. We shall need a lemma.

Lemma 13.5 *If a double sequence $\{B_{j,n}\}_{j,n\in\mathbb{N}}$ of sets from $\mathcal{S}$ fulfills (a), (b) and (c) from Definition 13.1, then there exists a double sequence $\{H_{m,p}\}_{m,p\in\mathbb{N}}$ of $\mathcal{S}$-measurable sets such that $B_{j,n} = \bigcup_{m=j}^{j+n-1} \bigcup_{p=j}^{j+n-1} H_{m,p}$ for each $j, n \in \mathbb{N}$.*

Proof of the Lemma Put $H_{m,m} = B_{m,1}$ for $m \in \mathbb{N}$, $H_{m,n} = B_{m,n-m+1} \setminus \big(B_{m,n-m} \cup B_{m+1,n-m}\big)$ if $m < n$ and $H_{m,n} = \emptyset$ if $m > n$. We shall proceed by induction with respect to n. For $n = 1$ the equality is obvious. Suppose now that $B_{j,n} = \bigcup_{m=j}^{j+n-1} \bigcup_{p=j}^{j+n-1} H_{m,p}$. Then

$$\bigcup_{m=j}^{j+n} \bigcup_{p=j}^{j+n} H_{m,p} = B_{j,n} \cup H_{j,j+n} \cup H_{j+1,j+n} \cup \ldots \cup H_{j+n-1,j+n} \cup H_{j+n,j+n} =$$

$$B_{j,n} \cup \big(B_{j,n+1} \setminus \big(B_{j,n} \cup B_{j+1,n}\big)\big) \cup \big(B_{j+1,n} \setminus \big(B_{j+1,n-1} \cup B_{j+2,n-1}\big)\big)$$
$$\cup \ldots \cup \big(B_{j+n-1,2} \setminus \big(B_{j+n-1,1} \cup B_{j+n,1}\big)\big) \cup B_{j+n,1} = B_{j,n+1}$$

The last equality follows from (a) and (c). Indeed,

$$B_{j+n-1,1} \subset B_{j+n-2,2} \cdots \subset B_{j+i,n-i} \subset \ldots \subset B_{j,n}$$

and

$$B_{j+n,1} \subset B_{j+n-1,2} \cdots \subset B_{j+i,n-i+1} \subset \ldots \subset B_{j+1,n} \subset B_{j,n+1}.$$

It is obvious that the union on the left-hand side of the last equality is included in $B_{j,n+1}$. Suppose that $x \in B_{j,n+1}$. If there exists $i \in \{0, 1, \ldots, n-1\}$ for which $x \in B_{j+i,n-i}$, then $x \in B_{j,n}$, which means that x belongs to the union on left-hand side. If $x \notin B_{j,n}$ and $i \in \{0, 1, \ldots, n-1\}$ is the smallest number for which $x \notin B_{j+i,n-i+1}$, then $x \in \big(B_{j+i,n-i+1} \setminus \big(B_{j+i,n-i} \cup B_{j+i+1,n-i}\big)\big)$, so x also belongs to the union on the left-side, at last if $x \in B_{j+n,1}$, then obviously x belongs to the union on the left-side. □

Proof of the Theorem Suppose that condition (F) is not fulfilled, so there exist a set $D \in \mathcal{S} \setminus \mathcal{I}$ and a double sequence $\{B_{j,n}\}_{j,n \in \mathbb{N}}$ of sets from $\mathcal{S}$ fulfilling (a), (b) and (c) from Definition 13.1 such that for each sequence $\{n_j\}_{j \in \mathbb{N}}$ we have $\lim_{j\to\infty} B_{j,n_j} \in \mathcal{I}$. Let $\{H_{m,p}\}_{m,p \in \mathbb{N}}$ be a double sequence associated with $\{B_{j,n}\}_{j,n \in \mathbb{N}}$ as in Lemma 13.5. Put $f_{m,p}(x) = \chi_{H_{m,p}}(x)$ for $m, p \in \mathbb{N}$. Let $\{m_k\}_{k \in \mathbb{N}}$, $\{n_k\}_{k \in \mathbb{N}}$ be sequences of positive integers such that $\lim_{k\to\infty} m_k = \lim_{k\to\infty} n_k = +\infty$. Observe that $H_{m_k,n_k} \subset B_{j_k,p_k}$, where $j_k = \min(m_k, n_k)$ and $p_k = |m_k - n_k| + 1$ for $k \in \mathbb{N}$. Hence $\limsup_{k\to\infty} H_{m_k,n_k} \subset \limsup_{k\to\infty} B_{j_k,n_k} \in \mathcal{I}$ since $\lim_{k\to\infty} j_k = +\infty$, so $\lim_{k\to\infty} f m_k, n_k = 0$ $\mathcal{I}$-a.e.

Simultaneously it is not difficult to see that

$$\{x \in X : \{f_{m,n}\}_{m,n \in \mathbb{N}} \text{ does not converge to } 0\} \supset D.$$

□

If $x \in [0, 1]$, $\mathcal{L}$ is a σ-algebra of Lebesgue measurable subsets of $[0, 1]$ and $\mathcal{N} \subset \mathcal{L}$ is a σ-ideal of null sets, then we obtain theorem of Sierpiński as an easy consequence of Theorem 13.4.

Let $X = [0, 1]$, $\mathcal{B}$ is a σ-algebra of subsets of $[0, 1]$ having the Baire property and $\mathcal{K} \subset \mathcal{B}$ is a σ-ideal of sets of the first category. Recall that the pair $(\mathcal{B}, \mathcal{K})$ does not fulfill the condition (E) (Theorem 1.20).

Theorem 13.6 ([Wag81]) *The pair $(\mathcal{B}, \mathcal{K})$ fulfills the condition (F).*

Proof Let $D \subset \mathcal{B} \setminus \mathcal{K}$ and let $\{B_{j,n}\}_{j,n \in \mathbb{N}}$ be a double sequence of sets from $\mathcal{B}$ fulfilling (a), (b) and (c) from Definition 13.1 (in fact the condition (c) is not used in the proof). If $I \subset (0, 1)$ is an open interval such that D is residual in I, then from (b) it follows that there exists $n_1 \in \mathbb{N}$ for which $B_{1,n_1} \cap I$ is of the second category, which means that there exists an interval $(a, b) \subset I$ in which B_{1,n_1} is residual. Suppose that we have defined B_{1,n_1}, B_{2,n_2} ..., B_{j,n_j}. Let $\{I_1, \ldots, I_{2^{j+1}}\}$ be the partition of I onto equal parts, then again by virtue of (b) there exists $n_{j+1} \in \mathbb{N}$ for which $B_{j+1,n_{j+1}} \cap I_p$ is of the second category (so residual in some subinterval of I_p) for $p \in \{1, 2, \ldots, 2^{j+1}\}$.

It is not difficult to see that $\limsup_{j\to\infty} B_{j,n_j} = \bigcap_{m=1}^{\infty} \bigcup_{j=m}^{\infty} B_{j,m_j}$ is residual in I. □

Theorem 13.7 ([Wag81]) *If $\{f_{m,n}\}_{m,n\in\mathbb{N}}$ is a double sequence of real-valued functions defined on $[0,1]$ having the Baire property such that for each sequences $\{m_k\}_{k\in\mathbb{N}}$ and $\{n_k\}_{k\in\mathbb{N}}$ of positive integers convergent to infinity the sequence $\{f_{m_k,n_k}\}_{k\in\mathbb{N}}$ converges $\mathcal{K}$-a.e., then the double sequence $\{f_{m,n}\}_{m,n\in\mathbb{N}}$ converges $\mathcal{K}$-a.e.*

Observe that the sequences $\{m_k\}_{k\in\mathbb{N}}$ and $\{n_k\}_{k\in\mathbb{N}}$ in the proof of Theorem 13.4 are far from being monotone. W. Sierpiński has proved that the situation is unavoidable in the case of measure.

Theorem 13.8 ([Sie50]) *There exists a sequence $\{f_{m,n}\}_{m,n\in\mathbb{N}}$ of continuous real-valued functions defined on $\mathbb{R}$ such that for each increasing sequences $\{m_k\}_{k\in\mathbb{N}}$ and $\{n_k\}_{k\in\mathbb{N}}$ of positive integers $\lim_{k\to\infty} f_{m_k,n_k} = 0$ a.e. on $\mathbb{R}$ and $\lambda\left(\left\{x\in\mathbb{R}\colon \{f_{m,n}\}_{m,n\in\mathbb{N}} \text{ does not converge to } 0\right\}\right) > 0$.*

Proof If $m < n$, put $f_{m,n}(x) = 0$ for $x \in \mathbb{R}$. If $m \geq n$, put

$$f_{m,n}(x) = \begin{cases} 0 \text{ for } x \leq \frac{m-n-1}{2^n} \text{ and } x \geq \frac{m-n+2}{2^n}, \\ 1 \text{ for } \frac{m-n}{2^n} \leq x \leq \frac{m-n-1}{2^n}, \\ \text{linear in the intervals (to obtain continuity) } \frac{m-n-1}{2^n} \leq x \leq \frac{m-n}{2^n} \\ \text{and } \frac{m-n+1}{2^n} \leq x \leq \frac{m-n+2}{2^n}. \end{cases}$$

Let $\{m_k\}_{k\in\mathbb{N}}$ and $\{n_k\}_{k\in\mathbb{N}}$ be increasing sequences of positive integers. If $E_{m,n} = \left[\frac{m-n-1}{2^n}, \frac{m-n+2}{2^n}\right]$, then $\lambda\left(\limsup_{k\to\infty} E_{m_k,n_k}\right) = 0$. Since $\lim_{k\to\infty} f_{m_k,n_k}(x) = 0$ for $x \notin \limsup_{k\to\infty} E_{m_k,n_k}$ we have $\lim_{k\to\infty} f_{m_k,n_k}(x) = 0$ a.e. If $x_0 \geq 0$, then for each $n \in \mathbb{N}$ there exists $m_n \geq n$ such that $\frac{m_n-n}{2^n} \leq x_0 \leq \frac{m_n-n+1}{2^n}$, so $f_{m_n,n}(x_0) = 1$ and $\{f_{m,n}(x_0)\}_{m,n\in\mathbb{N}}$ does not converge to 0. □

In the case of category situation is different.

Theorem 13.9 ([Wag81]) *If $\{f_{m,n}\}_{m,n\in\mathbb{N}}$ is a double sequence of real-valued functions defined on $[0,1]$ having the Baire property such that for each increasing sequences $\{m_k\}_{k\in\mathbb{N}}$ and $\{n_k\}_{k\in\mathbb{N}}$ of positive integers $\lim_{k\to\infty} f_{m_k,n_k} = 0$ $\mathcal{K}$-a.e., then $\lim_{m,n\to\infty} f_{n,m} = 0$ $\mathcal{K}$-a.e.*

Proof Suppose that the set $E = \{x \in X\colon \{f_{m,n}\}_{m,n\in\mathbb{N}} \text{ does not converge to } 0\}$ does not belong to $\mathcal{K}$. Since

$$E = \bigcup_{i=1}^{\infty} \bigcap_{M=1}^{\infty} \bigcup_{m=M}^{\infty} \bigcup_{n=M}^{\infty} \left\{x \in X\colon |f_{m,n}| > \frac{1}{i}\right\}$$

there exists $i_0 \in \mathbb{N}$ such that

$$D = \bigcap_{M=1}^{\infty} \bigcup_{m=M}^{\infty} \bigcup_{n=M}^{\infty} \left\{x \in X\colon |f_{m,n}| > \frac{1}{i_0}\right\} \notin \mathcal{K}.$$

Let $(a, b) \subset [0, 1]$ be an interval in which D is residual. Put $H_{m,n} = \{x \in X : |f_{m,n}| > \frac{1}{i_0}\}$ for $m, n \in \mathbb{N}$. There exist open sets $\{G_{m,n}\}_{m,n\in\mathbb{N}}$ included in (a, b) and sets of the first category $\{P_{m,n}\}_{m,n\in\mathbb{N}}$ such that $H_{m,n} = G_{m,n} \Delta P_{m,n}$ for $m, n \in \mathbb{N}$. Observe that for each positive integer M the open set $\bigcup_{m=M}^{\infty} \bigcup_{n=M}^{\infty} G_{m,n}$ is dense in (a, b). Let $\{(a_k, b_k)\}_{k\in\mathbb{N}}$ be the basis of the natural topology of (a, b). Choose m_1 and n_1 such that $G_{m_1,n_1} \cap (a_1, b_1) \neq \emptyset$. If we have chosen $m_1 < m_2 < \ldots < m_{k-1}$ and $n_1 < n_2 < \ldots < n_{k-1}$ in such a way that $G_{m_i,n_i} \cap (a_i, b_i) \neq \emptyset$ for $i \in \{1, 2, \ldots, k-1\}$, choose $m_k > m_{k-1}$ and $n_k > n_{k-1}$ such that $G_{m_k,n_k} \cap (a_k, b_k) \neq \emptyset$. Obviously $\limsup_{k\to\infty} G_{m_k,n_k}$ is residual in (a, b), so $\limsup_{k\to\infty} H_{m_k,n_k}$ is also residual in (a, b), which means that the sequence $\{f_{m,n}\}_{m,n\in\mathbb{N}}$ does not converge $\mathcal{K}$-a.e. to 0. □

Suppose now that $\mathcal{S} = 2^{[0,1]}$ and $\mathcal{I} \subset \mathcal{S}$ is a σ-ideal of at most denumerable subsets of $[0, 1]$. Assume the continuum hypothesis.

Theorem 13.10 ([Sie50]) *There exists a double sequence $\{f_{m,n}\}_{m,n\in\mathbb{N}}$ of real-valued functions defined on $[0, 1]$ such that for each sequences $\{m_k\}_{k\in\mathbb{N}}$ and $\{n_k\}_{k\in\mathbb{N}}$ of positive integers convergent to infinity the sequence $\{f_{m_k,n_k}\}_{k\in\mathbb{N}}$ converges to 0 except on a denumerable set and*

$$\{x \in X : \{f_{m,n}\}_{m,n\in\mathbb{N}} \textit{ does not converge to } 0\} = [0, 1].$$

Proof If $A = (a_1, a_2, \ldots)$ and $B = (b_1, b_2, \ldots)$ are sequences of positive integers, then we say that A precedes B ($A \prec B$) if there exists a positive integer i such that $a_k < b_k$ for $k \geq i$. Observe that if $\{A^{(i)}\}_{i\in\mathbb{N}}$ is a sequence of sequences of positive integers, $A^{(i)} = (a_1^{(i)}, a_2^{(i)}, \ldots)$ then there exists an increasing sequence B of positive integers for which $A^{(i)} \prec B$, $i \in \mathbb{N}$. It is sufficient to put $b_k = \sum_{i=1}^{k} \sum_{j=1}^{k} a_j^{(i)} + 1$. The set of all sequences of positive integers is of the power of continuum. Let $\{A_\alpha\}_{\alpha<\omega_1}$ be a transfinite sequence consisting of all sequences of positive integers. If $\alpha < \omega_1$, then there exists an increasing sequence of positive integers $B_\alpha = \{b_1^\alpha, b_2^\alpha, \ldots\}$ such that $A_\xi \prec B_\alpha$ for $\xi < \alpha$. If A_β does not precede B_α, then $\alpha \leq \beta$. Hence for each sequence A of positive integers the set $\{\alpha < \omega_1 : A \text{ does not precede } B_\alpha\}$ is at most denumerable. Let $\{x_\alpha\}_{\alpha<\omega_1}$ be a transfinite sequence consisting of all numbers from $[0, 1]$. Put

$$f_{m,n}(x_\alpha) = \begin{cases} 1 & \text{if } n = b_m^\alpha, \\ 0 & \text{if } n \neq b_m^\alpha. \end{cases}$$

Let $\{m_k\}_{k\in\mathbb{N}}$ and $\{n_k\}_{k\in\mathbb{N}}$ be sequences of positive integers convergent to infinity. Take a positive integer m. If m does not belong to the sequence $\{m_k\}_{k\in\mathbb{N}}$, put $a_m = m$. If there are some elements of $\{m_k\}_{k\in\mathbb{N}}$ equal to m, say $m_{j_1}, m_{j_2}, \ldots m_{j_s}$, put $a_m = \max(n_{j_1}, n_{j_2}, \ldots n_{j_s})$. Observe, that $a_{m_k} \geq n_k$ for $k \in \mathbb{N}$. Suppose that $x \in [0, 1]$ is such that $\{f_{m_k,n_k}(x)\}_{k\in\mathbb{N}}$ does not converge to 0. There exists an increasing sequence $\{k_i\}_{i\in\mathbb{N}}$ of positive integers such that $f_{m_{k_i},n_{k_i}}(x) = 1$ for $i \in \mathbb{N}$. Let $\alpha < \omega_1$ be such that $x = x_\alpha$. We have $a_{m_{k_i}} > n_{k_i} = b_{m_{k_i}}^\alpha$ for $i \in \mathbb{N}$. Hence the sequence $A = \{a_m\}_{m\in\mathbb{N}}$ does not precede B_α, so the set $\left\{x \in X : \{f_{m,n}\}_{m,n\in\mathbb{N}} \text{ does not converge to } 0\right\}$ is at most denumerable. Take now $x = x_\alpha$ for $\alpha < \omega_1$. Then $f_{m,b_m^\alpha}(x) = 1$ for $m \in \mathbb{N}$, so $\lim_{m,n\to\infty} f_{m,n}(x)$ cannot be 0, since $\{b_m^\alpha\}_{m\in\mathbb{N}}$ is increasing. □

Remark 13.11 Banach and Kuratowski [BK29] have proved that the pair ($2^{[0,1]}$, at most denumerable subsets of [0, 1]) does not fulfill the condition (E). From Theorems 13.10 and 13.4 it follows that this pair also does not fulfill the condition (F).

Remark 13.12 ([Sie50]) Let $X \subset [0, 1]$ be a Cantor set. Repeat the construction in the proof of Theorem 13.10 to obtain a double sequence $\{f_{m,n}\}_{m,n\in\mathbb{N}}$ of functions and put

$$g_{m,n}(x) = \begin{cases} f_{m,n}(x) & \text{if } x \in X \\ 0 & \text{if } x \in [0,1] \setminus X \end{cases}.$$

Then the double sequence $\{g_{m,n}\}_{m,n\in\mathbb{N}}$ consists of measurable functions having the Baire property, for each sequences $\{m_k\}_{k\in\mathbb{N}}$ and $\{n_k\}_{k\in\mathbb{N}}$ of positive integers convergent to infinity the sequence $\{f_{m_k,n_k}\}_{k\in\mathbb{N}}$ converges to 0 except on a denumerable set and

$$\left\{x \in X \colon \{g_{m,n}\}_{m,n\in\mathbb{N}} \text{ does not converge to } 0\right\} = X.$$

Hence both pairs

$$(\mathcal{L}, \text{at most denumerable subsets of } [0, 1])$$

and

$$(\mathcal{B}, \text{at most denumerable subsets of } [0, 1])$$

do not fulfill the condition (F).

Remark 13.13 ([Sie50]) The pair

$$(\text{Borel subsets of } [0, 1], \text{at most denumerable subsets of } [0, 1])$$

fulfills the condition (E).

We shall show that if $D \subset [0, 1]$ is a non-denumerable Borel set and $\{B_{j,n}\}_{j,n\in\mathbb{N}}$— a double sequence of of Borel sets such that $B_{j,n} \subset B_{j,n+1}$ for each $j, n \in \mathbb{N}$ and $\bigcup_{n=1}^{\infty} B_{j,n} = D$ for each $j \in \mathbb{N}$, there exist a perfect set $Q \subset [0, 1]$ and a sequence $\{n_j\}_{j\in\mathbb{N}}$ of positive integers such that $Q \subset \bigcap_{j=1}^{\infty} B_{j,n_j}$. Recall the Alexandrov-Hausdorff theorem (see [Kur66], p. 447): each non-denumerable Borel set includes a (non-empty) perfect set. Hence there exist $n_1 \in \mathbb{N}$ and a perfect set $P \subset B_{1,n_1}$. Let Q_0 and Q_1 be disjoint perfect non-empty sets included in P having diameters less than 2^{-1}. We have $Q_0 \cup Q_1 \subset B_{1,n_1}$. There exists $n_2 \in \mathbb{N}$ for which both sets $Q_0 \cap B_{2,n_2}$ and $Q_1 \cap B_{2,n_2}$ are non-denumerable, so it is possible to find disjoint non-empty perfect sets $Q_{0,0}$, $Q_{0,1}$, $Q_{1,0}$, $Q_{1,1}$ having diameters less than 3^{-1} such that $Q_{0,0} \cup Q_{0,1} \subset Q_0 \cap B_{2,n_2}$ and $Q_{1,0} \cup Q_{1,1} \subset Q_1 \cap B_{2,n_2}$. Proceeding in this way for each $j \in \mathbb{N}$ we can find $n_j \in \mathbb{N}$ and a family of sets $Q_{i_1,i_2,\ldots,i_j}$, where $i_1, i_2, \ldots, i_j \in \{0, 1\}$, such that

$$Q_{i_1,i_2,\dots,i_{j-1},0} \cup Q_{i_1,i_2,\dots,i_{j-1},1} \subset Q_{i_1,i_2,\dots,i_{j-1}} \subset B_{j_1,n_{j-1}},$$

$$P_j = \bigcup_{i=1}^{\infty} Q_{i_1,i_2,\dots,i_j} \subset B_{j,n_j},$$

all sets $Q_{i_1,i_2,\dots,i_j}$ are perfect and non-empty, and have diameters less than $(j+1)^{-1}$. Then $Q = \bigcap_{j=1}^{\infty} P_j$ is a perfect set fulfilling all requirements.

Chapter 14
Some Differences and Similarities Between Measure and Category

In this chapter we shall show that almost all numbers are normal in the scale of 2 while the set of normal numbers is of the first category. This is formulated in [Oxt80], Chap. 21, without proof.

We shall compare the convergence of the series $\sum_{n=1}^{\infty} a_n r_n(x)$ depending on the behaviour of the sequence $\{a_n\}_{n\in\mathbb{N}}$ where $\{r_n\}_{n\in\mathbb{N}}$ is the orthogonal and normalized sequence of Rademacher.

Theorems concerning the convergence and the divergence a.e. of the series $\sum_{n=1}^{\infty} a_n r_n(x)$ are due to Rademacher (Theorem 14.6) and Kolmogorov (Theorem 14.8).

The proof of Theorem 14.6 is based upon the book [S-N64] and the proof of Theorem 14.8—on a book [Sik59].

If a sequence $\{a_n\}_{n\in\mathbb{N}}$ converges, then also each subsequence $\{a_{n_m}\}_{m\in\mathbb{N}}$ of $\{a_n\}_{n\in\mathbb{N}}$ converges to the same limit. If $\{a_n\}_{n\in\mathbb{N}}$ is divergent, most of its subsequences diverges in the sense of measure and category. The proof of this fact is based upon the properties of homogeneous sets (called also tail-sets, see [Oxt80]).

Let $x = \sum_{n=1}^{\infty} d_n(x)2^{-n}$ be the non-terminating binary representation of $x \in (0, 1]$.

The number x is normal in the scale of 2 if $\lim\limits_{n\to\infty} \frac{1}{n}\sum_{k=1}^{n} d_k(x) = \frac{1}{2}$. The following theorem is well known.

Theorem 14.1 $\lambda\left(\left\{x \in (0, 1]\colon \lim_{n\to\infty} \frac{1}{n}\sum_{k=1}^{n} d_k(x) = \frac{1}{2}\right\}\right) = 1.$

Proof The result follows immediately from the strong law of large numbers. It suffices to treat $\{d_n(x)\}_{n\in\mathbb{N}}$ as a sequence of mutually independent random variables with a common distribution. □

Remark 14.2 One can find the strong law of large numbers in each book of probability theory. As concerns normal numbers, there is much stronger theorem: Almost all numbers are normal in any scale (see for example [HWr62]).

W. Wilczyński, *Convergence in Measure and in Category*, Lecture Notes in Mathematics 2394, https://doi.org/10.1007/978-3-032-23143-7_14

The situation is different for the category.

Theorem 14.3 *The set* $A = \left\{x \in (0, 1]\colon \lim_{n\to\infty} \frac{1}{n}\sum_{k=1}^{n} d_k(x) = \frac{1}{2}\right\}$ *is of the first category.*

Proof Suppose that it is not the case. Then there exists the interval $(a, b) \subset (0, 1]$ such that $(a, b) \setminus A \in \mathcal{K}$. Indeed, the set A is not only of the second category, but also has the Baire property, since each d_k is a function having the Baire property and

$$A = \bigcap_{p\in\mathbb{N}} \bigcup_{m\in\mathbb{N}} \bigcap_{n\geq m} \left\{x \in (0, 1]\colon \left|\frac{1}{n}\sum_{k=1}^{n} d_k(x) - \frac{1}{2}\right| < \frac{1}{p}\right\}.$$

Take $\varepsilon = \frac{1}{4}$. Put

$$A_m = \left\{x \in (a, b) \cap A]\colon \left|\frac{1}{n}\sum_{k=1}^{n} d_k(x) - \frac{1}{2}\right| < \varepsilon \text{ for each } n \geq m\right\}$$

for $m \in \mathbb{N}$. The sequence $\{A_m\}_{m\in\mathbb{N}}$ is ascending, consists of sets having the Baire property and $\bigcup_{m\in\mathbb{N}} A_m = (a, b) \cap A$. hence there exist $m_0 \in \mathbb{N}$ and the interval $(c, b) \subset (a, b)$ such that $(c, d) \setminus A_{m_0} \in \mathcal{K}$. If $x_0 \in (c, d)$, x_0 has the terminating binary representation, say $x_0 = \sum_{j=1}^{j_0} d_k(x)$, and $j_1 \in \mathbb{N}$ is sufficiently big, then $(x_0, x_0 + 2^{-j_1}) \subset (c, d)$ and $\frac{1}{n}\sum_{k=1}^{p} d_k(x) < \frac{1}{4}$ for each $x \in (x_0, x_0 + 2^{-j_1})$. It follows from the fact that $d_k(x) = 0$ for $k = j_0 + 1, \ldots, j_1$. Hence $(x_0, x_0 + 2^{-j_1}) \cap A_{m_0} = \emptyset$, so $(c, d) \setminus A_{m_0} \supset (x_0, x_0 + 2^{-j_1})$—a contradiction. □

Definition 14.4 The Rademacher orthogonal sequence consists of the following functions: $r_0(x) = 1$ and for $k \in \mathbb{N}$

$$r_k(x) = \begin{cases} 1 & \text{if } \frac{i-1}{2^k} < x \leq \frac{i}{2^k} \text{ and } i \text{ is an odd positive integer,} \\ -1 & \text{if } \frac{i-}{2^k} < x \leq \frac{i}{2^k} \text{ and } i \text{ is an even positive integer,} \end{cases}$$

where $i \in \{1, 2, \ldots, 2^k\}$.

Remark 14.5 All functions above are defined on the interval $(0, 1]$. One can find the concise definition: $r_0(x) = 1$ $r_k(x) = \operatorname{sgn}(\sin 2^k\pi x)$ for $k \in \mathbb{N}$. Here the domain equals $[0, 1]$ and each function r_k vanishes at points of the form $\frac{i}{2^k}$. It makes no difference in the text below. It is obvious that the Rademacher sequence is orthogonal and normalized. It is not complete. Indeed, from Definition 14.4 it follows immediately that for a function $f \in L^2(0, 1)$ having the symmetry property $f(x) = f(1 - x)$ we have $\int_0^1 r_k(x)f(x)dx = 0$ for $k \in \mathbb{N}$, moreover, if such a function fulfills the condition $\int_0^1 f(x)dx = 0$, then it is orthogonal to r_0. Observe also that $-r_n(x) = 2d_n(x) - 1$ for all x except a denumerable set and for $n \in \mathbb{N}$.

Theorem 14.6 ([Ra22]) *If the series* $\sum_{n=1}^{\infty} a_n^2$ *converges, then the series* $\sum_{k=0}^{\infty} a_k r_k(x)$ *converges almost everywhere on* $(0, 1]$.

Proof From the theorem of Riesz-Fischer it follows that the sequence of functions $s_n(x) = \sum_{k=0}^{n} a_k r_k(x)$ converges in the mean to the function $f \in L^2(0,1]$. If $(c,d) \subset (0,1)$, then $\left|\int_c^d (f(x) - s_n(x))dx\right|^2 \leq (d-c)\int_0^1 (f(x) - s_n(x))^2 dx$, as a result of Schwarz's inequality, so $\lim_{n\to\infty} \left|\int_c^d (f(x) - s_n(x))dx\right|^2 = 0$. Observe also that if $n > k$, then $\int_{\frac{i-1}{2^k}}^{\frac{i}{2^k}} r_n(x)dx = 0$ for each $i \in \{1, \dots, 2^k\}$. Hence, for $n > k$ we have

$$\int_{\frac{i-1}{2^k}}^{\frac{i}{2^k}} s_n(x)dx = \int_{\frac{i-1}{2^k}}^{\frac{i}{2^k}} \big(s_k(x) + a_{k+1}r_{k+1}(x) + \ldots + a_n r_n(x)\big)dx = \int_{\frac{i-1}{2^k}}^{\frac{i}{2^k}} s_k(x)dx,$$

so $\int_{\frac{i-1}{2^k}}^{\frac{i}{2^k}} f(x)dx = \lim_{n\to\infty} \int_{\frac{i-1}{2^k}}^{\frac{i}{2^k}} s_n(x)dx = \int_{\frac{i-1}{2^k}}^{\frac{i}{2^k}} s_k(x)dx$.

If $x_0 \in (0,1)$ is a point of continuity of each Rademacher function, then for each $k \in \mathbb{N}$ there exists $i_k \in \mathbb{N}$ such that $x_0 \in \left(\frac{i_k-1}{2^k}, \frac{i_k}{2^k}\right)$.

Let $F(x) = \int_0^x f(t)dt$. We have $s_k(x_0) = 2^k \cdot \int_{\frac{i-1}{2^k}}^{\frac{i}{2^k}} s_k(x)dx$, since s_k is constant on the interval $\left(\frac{i_k-1}{2^k}, \frac{i_k}{2^k}\right)$. Hence $s_k(x_0) = 2^k \int_{\frac{i-1}{2^k}}^{\frac{i}{2^k}} f(x)dx = F\left(\frac{i}{2^k}\right) - F\left(\frac{i-1}{2^k}\right)$ which means that

$$s_k(x_0) = \frac{F\left(\frac{i_k}{2^k}\right) - F\left(\frac{i_k-1}{2^k}\right)}{\frac{i_k}{2^k} - \frac{i_k-1}{2^k}}.$$

If the derivative $F'(x_0)$ exists, then $F'(x_0) = \lim_{k\to\infty} s_k(x_0)$, but the derivative $F'(x)$ exists almost everywhere, so $\{s_k(x)\}_{k\in\mathbb{N}}$ converges a.e. □

Remark 14.7 If $\{f_n\}_{n\in\mathbb{N}}$ is a sequence of a measurable functions defined on $[0,1]$, then the following statements are equivalent:

(i) $\sum_{n=1}^{\infty} f_n^2(t) < \infty$ for a.e. $t \in (0,1)$,
(ii) $\sum_{n=1}^{\infty} f_n(t) r_n(x)$ converges with respect to t for a.e. $x \in (0,1)$,
(iii) $\sum_{n=1}^{\infty} f_n(t) r_n(x)$ converges in measure for a.e. $x \in (0,1)$.

From Theorem 14.6 it easily follows that (i) implies (ii) and (iii). For the proof of the converse implication see [KS84].

Theorem 14.8 *If* $\sum_{n=1}^{\infty} a_n^2 = \infty$, *then the series* $\sum_{n=1}^{\infty} a_n r_n(x)$ *diverges a.e. on* $(0,1]$.

Proof Suppose that the series $\sum_{n=1}^{\infty} a_n r_n(x)$ converges on some set $A \subset [0,1]$ with $\lambda(A) > 0$. From the theorem of Egorov (Theorem 1.4) it follows that there exists a set $B \subset A$ such that $\lambda(B) > 0$ and the sequence $\{s_n\}_{n\in\mathbb{N}}$ of partial sums converges uniformly on B. Hence there exists $n_0 \in \mathbb{N}$ such that $|s_m(x) - s_n(x)| < 1$ when $m > n \geq n_0$ and $x \in B$. We have

$$\lambda(B) \geq \int_B |s_m(x) - s_n(x)|^2 dx = \lambda(B) \cdot \sum_{j=n+1}^{m} a_j^2 + 2 \cdot \sum_{n<j<k\leq m} a_j a_k \int_B r_j(x) r_k(x) dx$$

if $m > n \geq n_0$. Observe that $\{r_j(x)r_k(x)\}_{j<n;\, j,k\in\mathbb{N}}$ forms an orthogonal and normalized system and the numbers $\int_B r_j(x)r_k(x)dx$ are coefficients of the development of a characteristic function χ_B with respect to this system. From the Bessel's inequality it follows that the series $\sum_{j<k;\, j,k\in\mathbb{N}} \left(\int_B r_j(x)r_k(x)dx\right)^2$ converges, so there exists $n_1 \geq n_0$ such that

$$\sum_{n_1<j<k\leq m} \left(\int_B r_j(x)r_k(x)dx\right)^2 < \frac{1}{9}\,(\lambda(B))^2$$

for $m > n_1$. From the Cauchy's inequality we obtain

$$\left|\sum_{n_1<j<k\leq m} a_j a_k \int_B r_j(x)r_k(x)dx\right| \leq$$

$$\leq \sqrt{\sum_{n_1<j<k\leq m} \left(\int_B r_j(x)r_k(x)dx\right)^2} \cdot \sqrt{\sum_{n_1<j<k\leq m} (a_j a_k)^2} \leq \frac{1}{3}\lambda(B) \cdot \sum_{j=n_1+1}^{m} a_j^2$$

Hence $\lambda(B) \geq \lambda(B)(1 - \frac{2}{3}) \sum_{j=n_1+1}^{m} a_j^2$ and finally $\sum_{j=n_1+1}^{\infty} a_j^2 \leq 3$, which means that the series $\sum_{n=1}^{\infty} a_j^2$ converges—a contradiction. □

The behaviour of the series $\sum_{n=1}^{\infty} a_n r_n(x)$ with respect to category is different. It is well known that if the series $\sum_{n=1}^{\infty} |a_n| < \infty$, then also $\sum_{n=1}^{\infty} a_n^2 < \infty$, but not conversely. It will be more convenient to consider the series of the form $\sum_{n=1}^{\infty} a_n(2d_n(x) - 1)$.

Theorem 14.9 ([MS96]) *If the series $\sum_{n=1}^{\infty} a_n$ is not absolutely convergent, then the set $A = \left\{x \in [0,1]\colon \sum_{n=1}^{\infty} a_n(2d_n(x) - 1) \text{ converges}\right\}$ is of the first category.*

Proof Put

$$A_{m,k} = \left\{x \in [0,1]\colon \sum_{n=i}^{j} a_n(2d_n(x) - 1) < \frac{1}{k} \text{ for each } i, j \in \mathbb{N},\ j > i > m\right\}.$$

Observe that $A = \bigcap_{k=1}^{\infty} \bigcup_{m=1}^{\infty} A_{m,k}$. We shall show that $A_{m,k}$ is nowhere dense for each $m, k \in \mathbb{N}$, which is sufficient. Let $x_0 = \sum_{n=1}^{i} 2^{-n} d_n(x_0)$ be an arbitrary number having terminating binary representation with $i > m$. Since the series $\sum_{n=1}^{\infty} a_n$ is not absolutely convergent, there exist $j > i$ and numbers $c_{i+1}, c_{i+2}, \ldots, c_j \in \{0,1\}$ such that $\sum_{n=i+1}^{j} a_n(2c_n(x) - 1) = \sum_{n=i+1}^{j} |a_n| > \frac{1}{k}$. Put

$$B = \left\{x \in [0, 1] \colon d_n(x) = d_n(x_0) \text{ for } n = 1, 2, \ldots, i \text{ and } d_n(x) = c_n \text{ for } n = i + 1, \ldots, j\right\}.$$

It is clear that $B \cap A_{m,k} = \emptyset$ and that B is an interval (of length 2^{-j}). The choice of x_0 was arbitrary, so $A_{m,k}$ is nowhere dense. □

We conclude with the following obvious theorem.

Theorem 14.10 *If the series* $\sum_{n=1}^{\infty} a_n$ *is absolutely convergent, then* $\sum_{n=1}^{\infty} a_n(2d_n(x) - 1)$ *converges at each* $x \in [0, 1]$.

Let $\{a_n\}_{n\in\mathbb{N}}$ be a sequence of real numbers and $x = \sum_{n=1}^{\infty} d_n(x) \cdot 2^{-n}$ be again the non-terminating binary representation of $x \in (0, 1]$. With x we can associate the subsequence of $\{a_n(x)\}_{n\in\mathbb{N}}$ consisting of elements a_n for which $d_n(x) = 1$. It is clear that we can speak of almost all subsequences of $\{a_n(x)\}_{n\in\mathbb{N}}$ (with respect to measure and to category).

Definition 14.11 We shall say that the set $E \subset (0, 1]$ is homogeneous if it has the following property: if $x = \sum_{n=1}^{\infty} d_n(x) \cdot 2^{-n}$ is the non-terminating binary representation of x, then the point obtained by changing a finite numbers of digits also belongs to E.

Theorem 14.12 *A homogeneous measurable set has measure 0 or 1. A homogeneous set having the Baire property is of the first category or is residual in* $(0, 1]$.

Proof It follows immediately from Theorems 21.3 and 21.4 in [Oxt80]. □

Lemma 14.13 *If* $E \subset (0, 1]$ *is a measurable set such that* $\lambda(E) = 1$, *then there exists a measurable set* $C \subset E$ *such that* $\lambda(A) = 1$ *and if* $x \in C$, *then also* $1 - x \in C$.

Proof It suffices to take $C = E \cap \{1 - x \colon x \in E\}$. □

Exactly in the same way we obtain

Lemma 14.14 *If* $E \subset (0, 1]$ *is a set having the Baire property, residual in* $(0, 1]$, *then there exists a set* $C \subset E$ *having the Baire property, residual in* $(0, 1]$ *and such that if* $x \in C$, *then also* $1 - x \in C$.

Lemma 14.15 *If* $\{a_n\}_{n\in\mathbb{N}}$ *is an arbitrary sequence of real numbers, then the set* $C \in (0, 1]$ *corresponding to the convergent subsequence of* $\{a_n\}_{n\in\mathbb{N}}$ *is measurable and has the Baire property.*

Proof Take $x \in (0, 1]$. For $n \in \mathbb{N}$ let $x(n)$ be the smallest positive integer for which $\sum_{k=1}^{x(n)} d_k(x) = n$. It means that $d_{x(n)}(x) = 1$ and 1 appears between $d_1(x)$, ..., $d_{x(n)}(x)$ exactly n times. Put $f_n(x) = a_n(x)$ for $n \in \mathbb{N}$ and $x \in (0, 1]$. Then f_n is Baire one function for each $n \in \mathbb{N}$. Indeed, there is an increasing sequence $0 = t_0 < t_1 < \ldots < t_{m(n)} = 1$ such that f_n is constant on each interval $(t_{i-1}, t_i]$, $i \in \{1, 2, \ldots, m(n)\}$. Observe that $C = \{x \in (0, 1] \colon \{f_n(x)\}_{n\in\mathbb{N}} \text{ converges}\}$ and the set of points of convergence of Baire one (simultaneously Borel one) functions is a Borel set, so it is measurable and has the Baire property. □

Theorem 14.16 ([BP43]) *If a sequence* $\{a_n\}_{n\in\mathbb{N}}$ *is divergent, then almost all of subsequences are divergent.*

Proof Let $C \subset (0, 1]$ be the set of points corresponding to convergent subsequence of $\{a_n\}_{n\in\mathbb{N}}$. C is a measurable and homogeneous set, so $\lambda(C) = 0$ or $\lambda(C) = 1$. Suppose that $\lambda(C) = 1$. From Lemma 14.14 it follows that there exists $x_0 \in C$ which is not of the form $k \cdot 2^{-n}$ with $k, n \in \mathbb{N}$ such that $1 - x_0 \in C$. The subsequence corresponding to x_0 and $1 - x_0$ must not have common terms, which means that if $x_0 = \sum_{n=1}^{\infty} d_n(x_0) \cdot 2^{-n}$ and $1 - x_0 = \sum_{n=1}^{\infty} d_n(1 - x_0) \cdot 2^{-n}$, then $d_n(x_0) \neq d_n(1 - x_0)$ for each $n \in \mathbb{N}$. Moreover, each term of $\{a_n(x)\}_{n\in\mathbb{N}}$ belongs exactly to one of these subsequences. Since $x_0 \in C$ and $1 - x_0 \in C$, both subsequences are convergent, say to a' and a''. If $\{a_{n_m}\}_{m\in\mathbb{N}}$ is a convergent subsequence of $\{a_n\}_{n\in\mathbb{N}}$, then it has infinitely many common terms with subsequence corresponding to x_0 or to $(1 - x_0)$, so $\lim_{m\to\infty} a_{n_m} = a'$ or $\lim_{m\to\infty} a_{n_m} = a''$. Let C' and C'' be sets of points corresponding to subsequences convergent to a' and a'', respectively. Then $C' \cup C'' = C$, both C' and C'' are measurable and homogeneous, so one of them, say C', must be of measure 1. If x_0 above has been chosen from C' and $\{a_{n_m}\}_{m\in\mathbb{N}}$ is now an arbitrary subsequence of $\{a_n\}_{n\in\mathbb{N}}$, then it follows that either $\{a_{n_m}\}_{m\in\mathbb{N}}$ is subdivided into two convergent subsequences—the first consisting of terms of $\{a_{n_m}\}_{m\in\mathbb{N}}$ belonging to subsequence corresponding to x_0 and the second—to $(1 - x_0)$, both convergent to the same limit a', or $\{a_{n_m}\}_{m\in\mathbb{N}}$ differs from the subsequence corresponding to x_0 (or to $(1 - x_0)$) with finite number of terms, so in each case $\lim_{m\to\infty} a_{n_m} = a'$. Each subsequence of $\{a_n\}_{n\in\mathbb{N}}$ is convergent, hence $\{a_n\}_{n\in\mathbb{N}}$ is also convergent—a contradiction. Finally $\lambda(C) = 0$. □

Theorem 14.17 **([BP43])** *If a sequence* $\{a_n(x)\}_{n\in\mathbb{N}}$ *is divergent, then the set*

$$\{x \in (0, 1)\colon \text{ the sequence } \{a_n(x)\}_{n\in\mathbb{N}} \text{ corresponding to } x \text{ is convergent}\}$$

is of the first category.

Proof The proof is essentially the same making use with the second part of Theorem 14.12 and Lemma 14.14. □

Chapter 15
Miscellanea

In this chapter we present topics more or less connected with the main theme of the book. First is the Baire category dual to convergence in measure considered by G. Beer in [Bee81, Bee82] for upper semicontinuous functions. We shall show that this kind of convergence differs from that of Wagner. The second is the result of Goffman and Waterman [GW60], who simplified the theorem of Talalyan related to basic sequences in the family of measurable real-valued functions together with the Baire counterpart. The third is devoted to the classical theorem of Frechet describing the necessary and sufficient condition for the family of measurable functions to be compact (with respect to the convergence in measure). The fourth presents the theorem of Trebukova stating that using appropriately constructed automorphisms one can transform the convergent in measure sequence of functions onto a sequence convergent almost everywhere. Next there is proved that if a sequence of monotone functions converges in measure or in category, then it converges a.e. or $\mathcal{K}$-a.e., respectively. Further we describe the family of open and convex sets in the space of measurable or having the Baire property functions equipped with the suitable kind of convergence. At last we present two theorems of H. Auerbach and T. Šalát concerning divergent subseries of a divergent series of positive numbers.

G. Beer in [Bee81, Bee82] has considered a category dual for the convergence in measure in the family of upper semicontinuous functions defined on a compact metric space. Below we shall present his approach, enlarge the definition to contain the family of functions having the Baire property and show that this kind of convergence may lead to a different limit than the convergence in category. We shall restrict ourselves to real-valued bounded functions defined on $[0, 1]$.

Put $d\big((x_1, y_1), (x_2, y_2)\big) = \max(|x_1 - x_2|, |y_1 - y_2|)$ for $(x_1, y_1), (x_2, y_2) \in [0, 1] \times \mathbb{R}$. The metric d generates the Hausdorff distance in the family of closed (compact) subsets of $[0, 1] \times \mathbb{R}$ defined by $D(C_1, C_2) = \inf\{r : B(C_1, r) \supset C_2 \text{ and } B(C_2, r) \supset C_1\}$, where $B(C, r)$ is the union of closed balls whose centra run over C and the radius equals r.

W. Wilczyński, *Convergence in Measure and in Category*, Lecture Notes in Mathematics 2394, https://doi.org/10.1007/978-3-032-23143-7_15

In the sequel UC will denote the set of all bounded upper semicontinuous functions defined on $[0, 1]$. It is well known that if $f \in UC$, then its hypograph $\operatorname{hypo} f = \{(x, y) : x \in [0, 1] \text{ and } y \leq f(x)\}$ is closed. We shall identify a function $f \in UC$ with its hypograph.

Theorem 15.1 *If $f, g \in UC$, then $D(f, g) = D(\operatorname{hypo} f, \operatorname{hypo} g) < \infty$, so D is a metric in the family UC.*

Proof It follows immediately from the boundedness of considered functions. □

Theorem 15.2 *([Bee82]) A sequence $\{f_n\}_{n\in\mathbb{N}}$ of real-valued functions from UC converges to $f \in UC$ in the metric D if and only if*

(1) *for each $x \in [0, 1]$ and each sequence $\{x_n\}_{n\in\mathbb{N}}$ convergent to x $\limsup_{n\to\infty} f_n(x_n) \leq f(x)$,*
(2) *for each $x \in [0, 1]$ there exists a sequence $\{x_n\}_{n\in\mathbb{N}}$ convergent to x for which $\limsup_{n\to\infty} f_n(x_n) \geq f(x)$.*

Proof Suppose that (1) holds. Take $\varepsilon > 0$. For each $x \in [0, 1]$ there exists $\delta(x) \in (0, \varepsilon)$ and a positive integer $n(x)$ such that if $|x - z| < \delta(x)$ and $n \geq n(x)$, then $f_n(z) < f(x) + \varepsilon$. Choose $\{x_1, \ldots, x_k\} \subset [0, 1]$ for which $[0, 1] \subset \bigcup_{i=1}^{k} \{z : |x_i - z| < \delta(x_i)\}$. Let $n_0 = \max\{n(x_1), \ldots, n(x_k)\}$. If $x \in [0, 1]$ then there exists x_i for which $|x - x_i| < \delta(x_i) < \varepsilon$. By the definition of metric d in $[0, 1] \times \mathbb{R}$ we have $d\big((x, f_n(x)), (x_i, f_n(x) - \varepsilon)\big) = \varepsilon$. If $n \geq n_0$, then $(x_i, f_n(x) - \varepsilon) \in \operatorname{hypo} f$, so the graph of f_n is a subset of $B(\operatorname{hypo} f, \varepsilon)$ and $\operatorname{hypo} f_n \subset B(\operatorname{hypo} f, \varepsilon)$.

Suppose now that (2) holds. Again take $\varepsilon > 0$. For each $x \in [0, 1]$ there exists $\delta(x) \in (0, \frac{\varepsilon}{2})$ such that if $|x - z| < \delta(x)$, then $f(z) < f(x) + \frac{\varepsilon}{2}$. Choose $\{x_1, \ldots, x_k\}$ for which $[0, 1] \subset \bigcup_{i=1}^{k} \{z : |x_i - z| < \delta(x_i)\}$. From the assumption it follows that there exists n_0 such that for $n \geq n_0$ there exists $\{x_{n,1}, \ldots, x_{n,k}\} \subset [0, 1]$ such that for $i \in \{1, \ldots, k\}$ both $f(x_i) - f_n(x_{n,i}) < \frac{\varepsilon}{2}$ and $|x_{n,i} - x_i| < \delta(x_i)$. If $x \in [0, 1]$, then there exists x_i for which $|x - x_i| < \delta(x_i)$. Hence $|x - x_{n,i}| < \varepsilon$ and $f(x) < f_n(x_{n,i}) + \varepsilon$ for $n \geq n_0$. Thus $(x, f(x)) \in B(\operatorname{hypo} f_n, \varepsilon)$ for each $n \geq n_0$, so $\operatorname{hypo} f \subset B(\operatorname{hypo} f_n, \varepsilon)$. It means that (1) and (2) jointly imply that $\{f_n\}_{n\in\mathbb{N}}$ converges to f in the metric D.

Conversely suppose that $\lim_{n\to\infty} D(f_n, f) = 0$ and $\{x_n\}_{n\in\mathbb{N}}$ is a sequence such that $\lim_{n\to\infty} x_n = x$. From the definition of Hausdorff distance it follows that there exists a sequence $\{(z_n, y_n)\}_{n\in\mathbb{N}}$ of elements of $\operatorname{hypo} f$ such that $\lim_{n\to\infty} d\big((z_n, y_n), (x_n, f(x_n)\big) = 0$. Since f is upper semicontinuous at x and $\lim_{n\to\infty} z_n = x$ we have

$$\limsup_{n\to\infty} f_n(x_n) = \limsup_{n\to\infty} y_n \leq \limsup_{n\to\infty} f(z_n) \leq f(x)$$

which proves (1).

Similarly there exists a sequence $\{(x_n, y_n)\}_{n\in\mathbb{N}}$ such that $(x_n, y_n) \in \operatorname{hypo} f_n$ for each $n \in \mathbb{N}$ and $\lim_{n\to\infty} d\big((x_n, y_n), (x, f(x)\big) = 0$. Since $\lim_{n\to\infty} x_n = x$, $\lim_{n\to\infty} y_n = f(x)$ and $f_n(x_n) \geq y_n$ for each $n \in \mathbb{N}$, we obtain (2): $f(x) = \lim_{n\to\infty} y_n \leq \liminf_{n\to\infty} f_n(x_n)$. □

After [Bee81] we shall show that there is a subtle similarity between the convergence in category as described in Chap. 2 and the convergence of Beer.

Theorem 15.3 *([Bee81]) If $\{f_n\}_{n\in\mathbb{N}}$ is a sequence of real-valued continuous functions defined on* $[0, 1]$ *convergent to $f \in UC$ in the metric D, then there exists a dense set $G \in \mathcal{G}_\delta$ such that for each $x \in G$ there exists a subsequence $\{f_{n_m}\}_{m\in\mathbb{N}}$ for which* $\lim_{m\to\infty} f_{n_m}(x) = f(x)$.

Proof Take $r > 0$. We shall show that for each $k \in \mathbb{N}$ the closed set

$$A_{k,r} = \{x \in [0, 1] \colon f(x) - r \geq f_n(x) \text{ for each } n \geq k\}$$

is nowhere dense. Suppose that $x \in \operatorname{Int}(A_{k,r})$. From Theorem 15.2 it follows that there exists a sequence $\{x_n\}_{n\in\mathbb{N}}$ such that $\lim_{n\to\infty} x_n = x$ and $f(x) = \lim_{n\to\infty} f_n(x)$. So there exists $n_0 \in \mathbb{N}$ such that $f_n(x_n) > f(x) - \frac{r}{2}$, $f(x_n) < f(x) + \frac{r}{2}$ for each $n \geq n_0$. Hence

$$f_n(x_n) > f(x) - \frac{r}{2} > \left(f(x_n) - \frac{r}{2}\right) - \frac{r}{2} = f(x_n) - r$$

which contradicts the definition of $A_{k,r}$.

Let $E_{k,n} = A_{k,\frac{1}{n}}$ for $k, n \in \mathbb{N}$. Then $[0, 1] \setminus E_{k,n}$ is open and dense in $[0, 1]$, so $G = \bigcap_{k=1}^{\infty} \bigcap_{n=1}^{\infty} \left(([0, 1] \setminus E_{k,n}\right)$ is a dense $\mathcal{G}_\delta$ set. If $x \in G$ then there exists a subsequence $\{f_{n_k}\}_{k\in\mathbb{N}}$ of $\{f_n\}_{n\in\mathbb{N}}$ such that $f_{n_k}(x) > f(x) - \frac{1}{k}$ for each $k \in \mathbb{N}$, so $\liminf_{k\to\infty} f_{n_k}(x) \geq f(x)$. Again from Theorem 15.2 (1) $\limsup_{k\to\infty} f_{n_k}(x) \leq f(x)$ and finally $f(x) = \lim_{k\to\infty} f_{n_k}(x)$. □

Remark 15.4 The above kind of convergence has two unfavourable attributes. First: Theorem 15.3 works only for continuous functions. Indeed, let $f_n = \chi_{E_n}$, where $E_n = \{\frac{k}{2^n} \colon k = 0, 1, \ldots, 2^n\}$. Then the sequence $\{f_n\}_{n\in\mathbb{N}}$ consists of upper semi-continuous functions and converges to $f = \chi_{[0,1]}$ in the metric D, while $f(x)$ is a limit of some subsequence of $\{f_n\}_{n\in\mathbb{N}}$ if and only if x is a dyadic rational. The set of dyadic rationals is not a $\mathcal{G}_\delta$ set. Second: the subsequence depends on x and it can happen that there is not a single subsequence which is good for all points in a $\mathcal{G}_\delta$ dense set. For example [Bee82] let E_n be the same set as above and let $V_n = [0, 1] \cap \bigcup_{k=0}^{2^n} \left(\frac{k}{2^n} - 4^{-n}, \frac{k}{2^n} + 4^{-n}\right)$ and $f_n(x) = 4^n \cdot \operatorname{dist}(x, [0, 1] \setminus V_n)$, where $\operatorname{dist}(x, A) = \inf\{|x - z| \colon z \in A\}$. Each f_n is piecewise linear, $f_n(x) = 1$ for $x \in E_n$ and $f_n(x) = 0$ for $x \in [0, 1] \setminus V_n$. Put $g_n = 1 - f_n$. Both f_n and g_n are continuous. For each $x \in [0, 1]$ and $n \in \mathbb{N}$ there exists $x_n \in E_n$ such that $|x - x_n| < 2^{-n}$. We have $f_n(x_n) = 1$ and $g_n\left(x_n + 4^{-n}\right) = 1$. Since both sequences $\{x_n\}_{n\in\mathbb{N}}$ and $\{x_n + 4^{-n}\}_{n\in\mathbb{N}}$ converge to x, from Theorem 15.2 it follows that both sequences $\{f_n\}_{n\in\mathbb{N}}$ and $\{g_n\}_{n\in\mathbb{N}}$ converge to $\chi_{[0,1]}$. Suppose now that in Theorem 15.3 we can choose the same subsequence for all points of some dense $\mathcal{G}_\delta$ set. Then choosing first the subsequence from $\{f_n\}_{n\in\mathbb{N}}$ and next the second subsequence from $\{g_n\}_{n\in\mathbb{N}}$ we should obtain subsequences $\{f_{n_{m_p}}\}_{p\in\mathbb{N}}$ and $\{g_{n_{m_p}}\}_{p\in\mathbb{N}}$ convergent to $\chi_{[0,1]}$ on dense $\mathcal{G}_\delta$ sets G_1 and G_2, respectively. But then $\{f_{n_{m_p}} + g_{n_{m_p}}\}_{p\in\mathbb{N}}$ converges to 2 on the set

$G_1 \cap G_2$, which is also a dense $\mathcal{G}_\delta$ set. It is impossible because $f_n(x) + g_n(x) = 1$ for each $x \in [0, 1]$ and $n \in \mathbb{N}$.

Remark 15.5 Observe that if $f\colon [0, 1] \to \mathbb{R}$ is a bounded function having the Baire property, then there exists an upper semicontinuous function $\hat{f}\colon [0, 1] \to \mathbb{R}$ such that $\{x \in [0, 1]\colon f(x) \neq \hat{f}(x)\} \in \mathcal{K}$. Indeed, there exists a residual set $A \subset [0, 1]$ such that $f|_A$ is continuous. Put $\hat{f}(x) = \limsup_{z \in A, z \to x} f(z)$. Then $\hat{f} \in UC$ and $\{x \in [0, 1]\colon f(x) = \hat{f}(x)\} \supset A$. Observe also that if $g \in UC$ is such that $\{x \in [0, 1]\colon f(x) \neq g(x)\} \in \mathcal{K}$, then $\hat{f}(x) \leq g(x)$ for each $x \in [0, 1]$ (compare this argument with that of theorem 5.10).

Definition 15.6 We shall say that a sequence $\{f_n\}_{n \in \mathbb{N}}$ of real-valued function having the Baire property D-converges to a function $f\colon [0, 1] \to \mathbb{R}$ if and only if the sequence $\{\hat{f}_n\}_{n \in \mathbb{N}}$ converges to $\hat{f}$ in the metric D.

Theorem 15.7 *There exists a sequence $\{f_n\}_{n \in \mathbb{N}}$ of functions from UC (so obviously having the Baire property) convergent in category which is not D-convergent.*

Proof Put $f_{2n-1} = \chi_{E_n}$ and $f_{2n} = 2\chi_{E_n}$ for $n \in \mathbb{N}$ (here E_n is the same as in Remark 15.4). Then obviously $\{f_n\}_{n \in \mathbb{N}}$ converges in category to 0 and is not D-divergent because $D\,(\text{hypo}\, f_{2n-1}, \text{hypo}\, f_{2n}) = 1$ for each $k, n \in \mathbb{N}$. □

Theorem 15.8 *There exists a sequence $\{f_n\}_{n \in \mathbb{N}}$ of functions from UC which converges in the metric D but does not converge in category.*

Proof Put $f_{2n-1} = \chi_{E_n}$ and $f_{2n} = 2\chi_{E_n} - 1$ for $n \in \mathbb{N}$. Then $\{f_n\}_{n \in \mathbb{N}}$ converges to $\chi_{[0,1]}$ in the metric D and does not converge in category. □

Theorem 15.9 *There exists a sequence $\{f_n\}_{n \in \mathbb{N}}$ of functions from UC which converges in the metric D and in category to functions which differ on the set of the second category (in fact, on residual set).*

Proof Put $f_n = \chi_{E_n}$ for $n \in \mathbb{N}$. Then $\{f_n\}_{n \in \mathbb{N}}$ converges in the metric D to $\chi_{[0,1]}$ and in category to 0. □

Now we shall suppose that the measurable space $([0, 1], \mathcal{S})$ with σ-ideal $\mathcal{I}$ is either $([0, 1], \mathcal{L})$ equipped with the σ-ideal $\mathcal{N}$ or $([0, 1], \mathcal{B})$ equipped with $\mathcal{K}$, and the convergence of a sequence $\{f_n\}_{n \in \mathbb{N}}$ of $\mathcal{S}$-measurable real-valued functions defined on $[0, 1]$ is a convergence in measure or in category, respectively.

Definition 15.10 A sequence $\{f_n\}_{n \in \mathbb{N}}$ is called a basic sequence if for each $\mathcal{S}$-measurable function f there exists a sequence of finite linear combinations of functions from $\{f_n\}_{n \in \mathbb{N}}$ which converges $\mathcal{I}$-almost everywhere to f.

A. A. Talalyan in [Ta-n57] has observed that if $\{f_n\}_{n \in \mathbb{N}}$ is basic in the space of Lebesgue measurable functions, then if any finite number of functions is deleted from $\{f_n\}_{n \in \mathbb{N}}$, the remaining sequence is basic. His proof uses methods from the theory of orthonormal systems in L_2. C. Goffman and D. Waterman in [GW60] have found a very simple proof based upon the fact that the only linear functional on the space $\mathcal{M}$ of equivalence classes of measurable functions equals zero. Below we shall present their proof which works also in category case.

Theorem 15.11 *The dual space $\mathcal{M}'$ of $\mathcal{M}$ consists only of 0.*

Proof Let F be a continuous linear functional on $\mathcal{M}$. If $F(f) \neq 0$ for some f in $\mathcal{M}$, then either $F\left(f \cdot \chi_{[0,\frac{1}{2}]}\right) \neq 0$ or $F\left(f \cdot \chi_{[\frac{1}{2},1]}\right) \neq 0$. Denote by I_1 the interval $[0, \frac{1}{2}]$ if the first inequality holds and the interval $[\frac{1}{2}, 1]$ in the contrary case and put $f_1 = \frac{f \cdot \chi_{I_1}}{F(f \cdot \chi_{I_1})}$. In this way we obtain a descending sequence $\{I_n\}_{n \in \mathbb{N}}$ of intervals whose lengths converge to zero and a sequence $\{f_n\}_{n \in \mathbb{N}}$ of functions convergent $\mathcal{I}$-a.e. to zero such that $F(f_n) = 1$ for each $n \in \mathbb{N}$, a contradiction. □

Lemma 15.12 ([GW60]) *If the closure of the vector space generated by $f_1, f_2, \ldots$ is $\mathcal{M}$, and the closure of the vector space generated by $f_2, f_3, \ldots$ is $X \subset \mathcal{M}$, then $\mathcal{M}$ is the span of X and f_1.*

Proof If $f_1 \in X$, then the result is obvious, so suppose that $f_1 \notin X$. If $f \in \mathcal{M}$ then there exist a sequence $\{g_n\}_{n \in \mathbb{N}}$ of functions from X and a sequence $\{a_n\}_{n \in \mathbb{N}}$ of real numbers such that $\lim_{n\to\infty} (a_n \cdot f_1 + g_n) = f$ (the convergence $\mathcal{I}$-a.e.). Suppose that $\{a_n\}_{n \in \mathbb{N}}$ has a subsequence $\{a_{n_k}\}_{k \in \mathbb{N}}$ convergent to infinity. Then $\left\{f_1 + \frac{g_{n_k}}{a_{n_k}}\right\}_{k \in \mathbb{N}}$ would converge to 0, so $f_1 = -\lim_{k\to\infty} \frac{g_{n_k}}{a_{n_k}}$ would belong to X - a contradiction. Hence $\{a_n\}_{n \in \mathbb{N}}$ is bounded and has a convergent subsequence $\{a_{n_m}\}_{m \in \mathbb{N}}$. Let $a = \lim_{m\to\infty} a_{n_m}$. Observe that $a = \lim_{m\to\infty} \left(a_{n_m} \cdot f_1 + g_{n_m}\right) = f$ implies that $\{g_{n_m}\}_{m \in \mathbb{N}}$ converges to some $g \in X$ and thus $f = af_1 + g$. □

Lemma 15.13 ([GW60]) $X = \mathcal{M}$.

Proof Suppose that $X \neq \mathcal{M}$. By virtue of Lemma 15.12 there exists a continuous linear functional L on M which vanishes on X and such that $F(f_1) = 1$. This contradicts Theorem 15.11. □

Theorem 15.14 ([GW60]) *If $\{f_n\}_{n \in \mathbb{N}}$ is a basic sequence in $\mathcal{M}$, it remains basic if any finite subset is deleted.*

Proof From Lemma 15.13 it follows immediately that if $f_1, f_2, \ldots$ is a basic sequence in $\mathcal{M}$, then $f_2, f_3, \ldots$ is also a basic sequence and also for each $k \in \mathbb{N}$ the sequence $f_k, f_{k+1}, \ldots$ is basic. □

Remark 15.15 This is a suitable place to recall that both families of functions: measurable as well as having the Baire property possess basic sequences. By virtue of Theorems 5.4 and 5.10 and Weierstrass approximation theorem the sequence $\{1, x, x^2, \ldots\}$ is basic. Moreover, if $\mathcal{P}$ is the set of polynomials with rational coefficients, then for each measurable function $f : [0, 1] \to \mathbb{R}$ there exists a sequence $\{p_n\}_{n \in \mathbb{N}}$ of elements of $\mathcal{P}$ such that $f = \lim_{n\to\infty} p_n$ a.e. and similarly, for each function $f : [0, 1] \to \mathbb{R}$ having the Baire property there exists a sequence $\{p_n\}_{n \in \mathbb{N}}$ of elements of $\mathcal{P}$ such that $f = \lim_{n\to\infty} p_n$ $\mathcal{K}$-a.e.

Definition 15.16 Let $\{f_n\}_{n\in\mathbb{N}}$ be a basic sequence in the space $\mathcal{M}$ of equivalence classes of $\mathcal{S}$-measurable real-valued functions defined on $[0, 1]$. A series $\sum_{n=1}^{\infty} a_n f_n$ is called universal in a space $\mathcal{M}$ if for each $f \in \mathcal{M}$ there is a subsequence of partial sums of this series which converges to f except of a set belonging to the σ-ideal ($\mathcal{N}$ or $\mathcal{K}$, respectively).

Theorem 15.17 *If $\{f_n\}_{n\in\mathbb{N}}$ is a basic sequence in the space $\mathcal{M}$ of equivalence classes of Lebesgue measurable real-valued functions defined on* $[0, 1]$, *then there exists a sequence $\{a_n\}_{n\in\mathbb{N}}$ such that the series $\sum_{n=1}^{\infty} a_n f_n$ is universal in $\mathcal{M}$.*

Proof Recall (Theorem 2.11) that a space $\mathcal{M}$ is metrizable with the distance

$$d(f, g) = \int_0^1 \frac{|f(x) - g(x)|}{1 + |f(x) - g(x)|} \mathrm{d}x.$$

Let $\{g_n\}_{n\in\mathbb{N}}$ be a countable dense subset of $\mathcal{M}$ (for example the subset consisting of equivalence classes of functions generated by polynomials with rational coefficients) and let $\{\varepsilon_n\}_{n\in\mathbb{N}}$ be a sequence of positive numbers such that $\lim_{n\to\infty} \varepsilon_n = 0$. There exists a linear combination $S_{k_1} = \sum_{n=1}^{k_1} a_n f_n$ such that $d\left(g_1, S_{k_1}\right) < \varepsilon_1$. By virtue of Theorem 15.14 there exists a linear combination $\sum_{n=k_1+1}^{k_2} a_n f_n$ such that $d\Big(g_2 - S_{k_1}, \sum_{n=k_1+1}^{k_2} a_n f_n\Big) < \varepsilon_2$. Similarly there exists $\sum_{n=k_2+1}^{k_3} a_n f_n$ such that $d\Big(g_2 - S_{k_2}, \sum_{n=k_2+1}^{k_3} a_n f_n\Big) < \varepsilon_3$, where obviously $S_{k_2} = \sum_{n=1}^{k_2} a_n f_n$, and there exists: $\sum_{n=k_3+1}^{k_4} a_n f_n$ such that $d\Big(g_2 - S_{k_3}, \sum_{n=k_3+1}^{k_4} a_n f_n\Big) < \varepsilon_4$ and $\sum_{n=k_4+1}^{k_5} a_n f_n$ such that $d\Big(g_3 - S_{k_4}, \sum_{n=k_4+1}^{k_5} a_n f_n\Big) < \varepsilon_5$. We return to g_1 to approximate g_1, g_2, g_3, g_4 within $\varepsilon_6, \varepsilon_7, \varepsilon_8, \varepsilon_9$ and continue this process indefinitely. It is clear that for each $f \in \mathcal{M}$ there exists a subsequence of the sequence of partial sums of $\sum_{n=1}^{\infty} a_n f_n$ which converges a.e. to f. $\square$

The situation when $\mathcal{S} = \mathcal{B}$ is different. Now we shall show that a basic sequence $\{x^n\}_{n\in\mathbb{N}}$ forms a universal series in the family of equivalence classes of functions having the Baire property on the interval $[\frac{1}{2}, 1]$. The change of the domain is to avoid the fact that most of polynomials under consideration vanishes at 0. We shall need the following theorem of Julius Pál.

Theorem 15.18 *([Pal14]) If $f : [0, 1] \to \mathbb{R}$ is a continuous function and $a_0 = f(0)$, then for each $\varepsilon > 0$ and $m \in \mathbb{N}$ there exists a polynomial $p(x)$ with initial coefficients a_0, $a_1 = a_2 = \cdots = a_m = 0$ such that $|f(x) - p(x)| < \varepsilon$ for each $x \in [0, 1]$.*

Proof There exists $\delta > 0$ such that $|f(x) - a_0| < \frac{\varepsilon}{3}$ for $x \in [0, \delta]$. Let

$$g(x) = \begin{cases} \frac{f(x)-a_0}{x^{m+1}} & \text{for } x \in [\delta, 1] \\ 0 & \text{for } x = 0 \\ \text{linear and continuous} & \text{on } [0, \delta] \end{cases}$$

The function g is continuous on $[0, 1]$, so by virtue of the Weierstrass approximation theorem there exists a polynomial $q(x)$ such that

$$|g(x) - q(x)| < \frac{\varepsilon}{3} \quad \text{for each } x \in [0, 1].$$

Put $p(x) = a_0 + x^{m+1}q(x)$. We have $|f(x) - p(x)| < \frac{\varepsilon}{3}x^{m+1}$ for each $x \in [\delta, 1]$. For $x \in [0, \delta]$ we have $|g(x) - q(x)| < \frac{\varepsilon}{3}$ and $|g(x)| \leq g(\delta)$, so $|q(x)| \leq \frac{\varepsilon}{3} + g(\delta)$. Finally

$$\begin{aligned} |f(x) - p(x)| &< |f(x) - a_0| + x^{m+1}|q(x)| < \frac{\varepsilon}{3} + \delta^{m+1}|q(x) \\ &\leq \frac{\varepsilon}{3} + \frac{\varepsilon}{3} + \delta^{m+1}|g(\delta)| \\ &= \frac{2}{3}\varepsilon + |f(\delta) - g(\delta)| < \varepsilon \end{aligned}$$

for $x \in [0, \delta]$. The polynomial $p(x)$ has initial coefficients $a_0, a_1 = a_2 = \cdots = a_m = 0$, as prescribed. □

Corollary 15.19 *If $f\colon [\frac{1}{2}, 1] \to \mathbb{R}$ is a continuous function, then for each $\varepsilon > 0$ and each $m \in \mathbb{N}$ there exists a polynomial $p(x)$ with rational coefficients and initial coefficients $a_0 = a_1 = a_2 = \cdots = a_m = 0$ such that $|f(x) - p(x)| < \varepsilon$ for each $x \in [\frac{1}{2}, 1]$.*

Proof Put

$$F(x) = \begin{cases} f(x) & \text{for } x \in \left[\frac{1}{2}, 1\right] \\ 0 & \text{for } x = 0 \\ \text{linear and continuous} & \text{on } \left[0, \frac{1}{2}\right] \end{cases}$$

and apply Theorem 15.18. □

Theorem 15.20 *There exists a universal series $\sum_{n=1}^{\infty} a_n f_n$ in the space $\mathcal{M}$ of equivalence classes of functions having the Baire property defined on $[\frac{1}{2}, 1]$.*

Proof Observe that $f\colon [\frac{1}{2}, 1] \to \mathbb{R}$ is a function having the Baire property and $\{g_n\}_{n\in\mathbb{N}}$ is a sequence of continuous functions such that $\lim_{n\to\infty} g_n = f$ $\mathcal{K}$-a.e. on $[\frac{1}{2}, 1]$, and $\{p_n\}_{n\in\mathbb{N}}$ is a sequence of polynomials fulfilling the condition $|p_n(x) - g_n(x)| < \frac{1}{n}$ for $x \in [\frac{1}{2}, 1]$ and $n \in \mathbb{N}$, then $\lim_{n\to\infty} p_n = f$ $\mathcal{K}$-a.e.

Now it is sufficient to repeat the proof of Theorem 15.17 with $f_n(x) = x^n$ for $n \in \mathbb{N}$, a sequence $\{\varepsilon_n\}_{n\in\mathbb{N}}$ convergent to 0 and $\{g_n\}_{n\in\mathbb{N}}$ – a sequence of polynomials with rationals coefficients. □

Remark 15.21 More information concerning universal families in the real analysis, the complex analysis and the functional analysis setting one can find in [G-E99].

The classical theorem of Ascoli and Arzelá characterizes relatively compact families of continuous functions with respect to the uniform convergence. Frechet in [Frt27] has proved the theorem giving the characterization of relatively compact families of measurable functions with respect to the convergence in measure (or what amounts to the same, with respect to the convergence almost everywhere). Observe first that if $f\colon [0,1] \to \mathbb{R}$ is a measurable function, then for each $\varepsilon > 0$ there exists $p \in \mathbb{N}$ such that $\lambda(\{x \in [0,1]\colon |f(x)| \leq p\}) > 1 - \varepsilon$. Indeed, $\bigcap_{p=1}^{\infty}\{x \in [0,1]\colon |f(x)| > p\} = \varnothing$, so $\lim_{p\to\infty} \lambda(\{x \in [0,1]\colon |f(x)| > p\}) = 0$. Observe also that if $f\colon [0,1] \to \mathbb{R}$ is measurable, then for each $\varepsilon > 0$ there exists a closed set $F \subset [0,1]$ such that $\lambda(F) > 1 - \varepsilon$ and $f|_F$ is continuous (in fact, uniformly continuous). It is a consequence of Lusin theorem.

Hence with each measurable function $f\colon [0,1] \to \mathbb{R}$ and each pair of positive numbers ε, η we can associate the smallest integer $p(f,\varepsilon)$ and the closed set $F_f \subset [0,1]$ such that $\lambda(F_f) > 1 - \varepsilon$, $F_f \subset \{x \in [0,1]\colon |f(x)| < p(f,\varepsilon)\}$ and $f|_{F_f}$ is continuous, and the smallest integer $k(f,\varepsilon,\eta)$ such that if $\{I_1,\ldots,I_k\}$ is a subdivision of $[0,1]$ onto $k(f,\varepsilon,\eta)$ equal closed subintervals, then $\operatorname{osc}(f, I_i \cap F_f) \leq \eta$ for each $i \in \{1,\ldots,k(f,\varepsilon,\eta)\}$.

Definition 15.22 Let $\mathcal{M}$ be a family of measurable real-valued functions defined on $[0,1]$. We shall say that $\mathcal{M}$ is compact in measure if and only if for each sequence $\{f_n\}_{n\in\mathbb{N}}$ of functions from $\mathcal{M}$ there exists an increasing sequence $\{n_m\}_{m\in\mathbb{N}}$ of positive integers such that $\{f_{n_m}\}_{m\in\mathbb{N}}$ converges in measure (not necessarily to a function belonging to $\mathcal{M}$).

Definition 15.23 The family $\mathcal{M}$ is called uniformly quasi-bounded if and only if $\sup\{p(f,\varepsilon)\colon f \in \mathcal{M}\} < \infty$ for each $\varepsilon > 0$.

Definition 15.24 The family $\mathcal{M}$ is called uniformly measurable if and only if $\sup\{k(f,\varepsilon,\eta)\colon f \in \mathcal{M}\} < \infty$ for each $\varepsilon > 0$ and $\eta > 0$.

Remark 15.25 If a family $\mathcal{M}$ is uniformly measurable, then for each $\varepsilon > 0$ and $\eta > 0$ there exists $k \in \mathbb{N}$ such that if $\{I_1,\ldots,I_k\}$ is a family of non-overlapping closed intervals of equal length and $[0,1] = \bigcup_{i=1}^{k} I_i$, then for each $f \in \mathcal{M}$ there exists a closed set $F_f \subset [0,1]$, $\lambda(F_f) > 1 - \varepsilon$ such that $\operatorname{osc}(f, I_i \cap F_f) \leq \eta$ for $i \in \{1,\ldots,k\}$. The number k can be the least common multiple of $\{k(f,\varepsilon,\eta)\colon f \in \mathcal{M}\}$ (observe that the set $\{k(f,\varepsilon,\eta)\colon f \in \mathcal{M}\}$ is finite).

Lemma 15.26 *([Frt27]) If $\{f_n\}_{n\in\mathbb{N}}$ is a sequence of functions from $\mathcal{M}$ convergent in measure to some function $f\colon [0,1] \to \mathbb{R}$, then the set $\mathcal{M}_0 = \{f_n\colon n \in \mathbb{N}\}$ is uniformly quasi-bounded and uniformly measurable.*

Proof Take $\varepsilon > 0$ and $\eta > 0$. Take into account numbers $p(f,\frac{\varepsilon}{2})$ and $k(f,\frac{\varepsilon}{2},\frac{\eta}{3})$ Let F_f be a closed set such that $\lambda(F_f) > 1 - \frac{\varepsilon}{2}$ and $f|_{F_f}$ is continuous. There exists $n_0 \in \mathbb{N}$ such that $\lambda\left(\{x \in [0,1]\colon |f_n(x) - f(x)| < \frac{\eta}{3}\}\right) > 1 - \frac{\varepsilon}{2}$ for $n \geq n_0$. Let $F_n = F_f \cap \{x \in [0,1]\colon |f_n(x) - f(x)| < \frac{\eta}{3}\}$. Obviously $\lambda(F_n) > 1 - \varepsilon$ for $n \geq n_0$. If $x \in F_n$, then

$$|f_n(x)| \le |f(x)| + |f_n(x) - f(x)| < n\left(f, \frac{\varepsilon}{2}\right) + \frac{\eta}{3},$$

if $x_1, x_2 \in I_i \cap F_n$ for $i \in \{1, \ldots, k(f, \frac{\varepsilon}{2}, \frac{\eta}{3})\}$, then

$$\begin{aligned}|f_n(x_1) - f_n(x_2)| &< |f_n(x_1) - f(x_1)| + |f(x_1) - f(x_2)| + |f(x_2) - f_n(x_2)|\\ &< 3 \cdot \frac{\eta}{3} = \eta,\end{aligned}$$

all for $n \ge n_0$. Hence the family $\{f_{n_0}, f_{n_0+1}, \ldots\}$ is uniformly quasi-bounded and uniformly measurable. A finite family of measurable functions from Φ has always these properties and the union of two uniformly quasi-bounded and uniformly measurable families is also quasi-bounded and uniformly measurable. □

Theorem 15.27 ([Frt27]) *The family $\mathcal{M}$ is compact if and only if it is uniformly quasi-bounded and uniformly measurable.*

Proof Necessity. Suppose first that $\mathcal{M}$ is not uniformly quasi-bounded. Then there exists $\varepsilon_0 > 0$ such that for each $n \in \mathbb{N}$ there exists $f_n \in \mathcal{M}$ such that for each measurable set $F \subset [0, 1]$ if $\lambda(F) > 1 - \varepsilon_0$, then $\sup\{f_n(x) \colon x \in F\} > n$. Hence $\lambda(\{x \in [0, 1] \colon |f_n(x)| > n\}) \ge \varepsilon_0$ for each $n \in \mathbb{N}$. Indeed, suppose that $\lambda(\{x \in [0, 1] \colon |f_n(x)| > n\}) < \varepsilon_0$ for some $n \in \mathbb{N}$. Then for $F = [0, 1] \setminus \{x \in [0, 1] \colon |f_n(x)| > n\}$ we have $\lambda(F) > 1 - \varepsilon_0$ and $\sup\{f_n(x) \colon x \in F\} \le n$, a contradiction. There is no any subsequence of $\{f_n\}_{n\in\mathbb{N}}$ which converges in measure, so $\mathcal{M}$ is not compact in this case.

Now suppose that $\mathcal{M}$ is not uniformly measurable. It is not difficult to observe that there exist $\varepsilon_0 > 0$ and $\eta_0 > 0$ such that for each $n \in \mathbb{N}$ there exists $f_n \in \mathcal{M}$ for which $k(f_n, \varepsilon_0, \eta_0) \ge n$. Hence neither $\{f_n \colon n \in \mathbb{N}\}$, nor any $\{f_{n_m} \colon m \in \mathbb{N}\}$ is uniformly measurable, so by virtue of Lemma 10.4 $\{f_n\}_{n\in\mathbb{N}}$ also has not any subsequence convergent in measure and $\mathcal{M}$ is not compact.

Sufficiency. Let $\{f_n\}_{n\in\mathbb{N}}$ be a sequence of functions from $\mathcal{M}$. Take $\varepsilon > 0$ and consider p and k such that for each $n \in \mathbb{N}$ there exists a set $E_n \subset [0, 1]$ such that $\lambda(E_n) < \frac{\varepsilon}{5}$ and $f_n|_{E_n}$ is continuous and if $\{I_1, \ldots, I_k\}$ is a subdivision of $[0, 1]$ into k equal parts, then $\operatorname{osc}(f_n, I_n \setminus E_n) \le \frac{\varepsilon}{5}$ for each $n \in \mathbb{N}$ and $i \in \{1, 2, \ldots, k\}$. The existence of p and k follows from Definition 15.23, Definition 15.24 and Remark 15.25. If $y_i^n = \sup\{f_n(x) \colon x \in I_n \setminus E_n\}$, then $|y_i^n| \le p$ for each $n \in \mathbb{N}$ and $i \in \{1, 2, \ldots, k\}$. The sequence $q_n = \{y_1^n, y_2^n, \ldots, y_k^n\}_{n\in\mathbb{N}}$ of points in k-dimensional space is bounded, so there exists a subsequence $\{q_{n_m}\}_{m\in\mathbb{N}}$ such that $\rho(q_{n_m}, q_{n_r}) < \frac{\varepsilon}{3}$ for each $m, r \in \mathbb{N}$. So

$$|f_{n_m}(x) - f_{n_r}(x)| \le |f_{n_m}(x) - y_i^{n_m}| + |y_i^{n_m} - y_i^{n_r}| + |y_i^{n_r} - f_{n_r}(x)| \le 3 \cdot \frac{\varepsilon}{3} = \varepsilon$$

for each $x \in I_i \setminus (E_{n_r} \cup E_{n_m})$ and obviously $|f_{n_m}(x) - f_{n_r}(x)| \le \varepsilon$ for each $x \in [0, 1] \setminus (E_{n_r} \cup E_{n_m})$. Hence $\lambda\left(\{x \colon |f_{n_m}(x) - f_{n_r}(x)| > \varepsilon\}\right) \le \frac{2}{3}\varepsilon < \varepsilon$ and $\| f_{n_m} - f_{n_r} \| \le \varepsilon$ for each $n_m, n_r \in \mathbb{N}$. That means that for each $\varepsilon > 0$ one can find a subsequence of a given sequence such that the distance (compare Chap. 2) between arbitrary two elements of this subsequence is less than ε. Let $\left\{\{f_n^t\}_{n\in\mathbb{N}}\right\}_{t\in\mathbb{N}}$ be a sequence of subsequences of $\{f_n\}_{n\in\mathbb{N}}$ such that $\{f_n^1\}_{n\in\mathbb{N}}$ is a subsequence of $\{f_n\}_{n\in\mathbb{N}}$ and $\{f_n^{t+1}\}_{n\in\mathbb{N}}$ is a subsequence of $\{f_n^t\}_{n\in\mathbb{N}}$ such that $\| f_n^t - f_m^t \| < \frac{1}{t}$ for each $n, m, t \in \mathbb{N}$. A diagonal sequence $\{f_t^t\}_{t\in\mathbb{N}}$ fulfills the Cauchy condition in measure with respect to $\| \cdot \|$, so it is convergent in measure (see Theorem 3.5). □

Similar question concerning the convergence with respect to a σ-ideal is studied in [HWa91].

Now we shall present a result concerning the convergence in measure for which it is difficult, if not impossible, to formulate the category dual. This result is due to Trebukova. Our proof differs from the original one.

Definition 15.28 A one-to-one measurable map q of a Lebesgue measure space $([0, 1], \mathcal{L}, \lambda)$ onto itself such that q^{-1} is also measurable and both q and q^{-1} preserve measure is called an automorphism.

Theorem 15.29 *([Tre60]) If a sequence $\{f_n\}_{n\in\mathbb{N}}$ of real-valued measurable functions defined on $[0, 1]$ converges in measure to 0, then there exists a sequence $\{q_n\}_{n\in\mathbb{N}}$ of automorphisms such that the sequence $\{f_n \circ q_n^{-1}\}_{n\in\mathbb{N}}$ converges a.e. to 0.*

Proof We shall need the following lemma.

Lemma 15.30 *If A and B are disjoint measurable subsets of $[0, 1]$ such that $\lambda(A \cup B) = 1$, then there exist an automorphism q such that $q(B) \subset (0, \lambda(B)]$.*

Proof of the Lemma Put $h_1(x) = \int_0^x \chi_{B\cap\Phi(B)}(t)dt$ for $x \in B \cap \Phi(B)$, where $\Phi(B)$ is the set of all points of density of B. Then h_1 is strictly increasing on the set $B \cap \Phi(B)$ and $h_1\,(B \cap \Phi(B)) = (0, \lambda(B))$. Moreover h_1 preserves measure of subsets of $B \cap \Phi(B)$. Indeed, if $E \subset B \cap \Phi(B)$ is measurable, then $h_1(E)$ is also measurable and $\lambda(h_1(E)) = \lambda(E)$, since $h_1'(x) = 1$ for each $x \in B \cap \Phi(B)$ (compare Lemmas 7.1 and 7.4, [BBT97], pp. 282 and 285). Thus h_1 is one-to-one correspondence between $B \cap \Phi(B)$ and $(0, \lambda(B))$, both h_1 and h_1^{-1} are measurable and preserve measure. If we put $h_2(x) = \lambda(B) + \int_0^x \chi_{A\cap\Phi(A)}(t)dt$ for $x \in A \cap \Phi(A)$, then similarly as above we obtain h_2 is one-to-one strictly increasing correspondence between $A \cap \Phi(A)$ and $(\lambda(B), 1)$, both h_2 and h_2^{-1} are measurable and preserve measure. Hence the function

$$h_3(x) = \begin{cases} h_1(x) & \text{if } x \in B \cap \Phi(B) \\ h_2(x) & \text{if } x \in A \cap \Phi(A) \end{cases}$$

is one-to-one correspondence between $(A \cap \Phi(A)) \cup (B \cap \Phi(B))$ and $(0, 1) \setminus \{\lambda(B)\}$, both h_3 and h_3^{-1} are measurable and preserve measure.

To finish the proof of the lemma we should add up the set $[0, 1] \setminus ((A \cap \Phi(A)) \cup (B \cap \Phi(B)))$ to the domain of h_3 and the set $\{0, \lambda(B), 1\}$ to the range. Put $D = (0, 1) \setminus ((A \cap \Phi(A)) \cup (B \cap \Phi(B)))$. Observe That $D \neq \varnothing$, because both sets are open in density topology, $(0, 1)$ is also open and connected in the density topology. Consider two cases:

First: $\operatorname{card} D = n_0 < \aleph_0$, so $D = \{z_1, z_2, \ldots, z_{n_0}\}$. Choose $E = \{x_0, x_1, \ldots, x_{n_0}, \ldots\} \subset B \cap \Phi(B)$. Put

$$h_4(x) = \begin{cases} \lambda(B) & \text{if } x = x_0 \\ x_i & \text{if } x = z_i,\ i \in \{1, 2, \ldots n_0\} \\ x_{i+n_0} & \text{if } x = x_i,\ i \in \mathbb{N} \end{cases}.$$

Then h_4 is one-to-one correspondence between $D \cup E$ and $E \cup \{\lambda(B)\}$. Next put

$$q(x) = \begin{cases} h_1(x) & \text{if } x \in (B \cap \Phi(B)) \setminus E \\ h_1(h_4(x)) & \text{if } x \in D \cup E \\ h_2(x) & \text{if } x \in A \cap \Phi(A) \\ 0 & \text{if } x = 0 \\ 1 & \text{if } x = 1 \end{cases}.$$

Second: $\operatorname{card} D \geq \aleph_0$. Choose disjoint subsets E_1, E_2 of $B \cap \Phi(B)$ such that $\lambda(E_1) = \lambda(E_2) = 0$, $\operatorname{card} E_1 = \operatorname{card} E_2 = \operatorname{card} D$ and a point $x_0 \in (B \cap \Phi(B)) \setminus (E_1 \cup E_2)$. Then let $h_4\colon E_1 \cup E_2 \to E_1$ and $h_5\colon D \to E_2$ be one-to-one correspondences. Put

$$q(x) = \begin{cases} h_1(x) & \text{if } x \in (B \cap \Phi(B)) \setminus (E_1 \cup E_2) \\ h_1(h_4(x)) & \text{if } x \in E_1 \cup E_2 \\ h_1(h_5(x)) & \text{if } x \in D \\ h_2(x) & \text{if } x \in A \cap \Phi(A) \\ 0 & \text{if } x = 0 \\ 1 & \text{if } x = 1 \end{cases}.$$

Then q in both cases is a required automorphism. □

Let's return to the proof of the theorem. If $\{f_n\}_{n\in\mathbb{N}}$ converges in measure to 0, then for each $k \in \mathbb{N}$ there exists $n_k \in \mathbb{N}$ such that

$$\lambda\left(\left\{x \in [0, 1] \colon |f_n(x)| \geq \frac{1}{k}\right\}\right) < \frac{1}{k} \quad \text{for } n \geq n_k.$$

We can and shall suppose that the sequence $\{n_k\}_{k\in\mathbb{N}}$ is increasing. Put

$$B_n = \left\{x \in [0, 1] \colon |f_n(x)| \geq \frac{1}{k}\right\} \quad \text{and} \quad A_n = \left\{x \in [0, 1] \colon |f_n(x)| < \frac{1}{k}\right\}$$

for $n \in \{n_k, \ldots, n_{k+1} - 1\}$, $k \in \mathbb{N}$. From the lemma it follows that for each $n \in \mathbb{N}$ there exists an automorphism q_n such that $q_n(B_n) \subset (0, \lambda(B_n)]$. Then the sequence $\{f_n \circ q_n^{-1}\}_{n\in\mathbb{N}}$ converges to 0 a.e. □

We shall show that for a sequence of real-valued monotone functions the convergence in measure is equivalent to the convergence a.e. and that for the category the situation is similar.

Theorem 15.31 *If* $\{f_n\}_{n\in\mathbb{N}}$ *is a sequence of real-valued nondecreasing functions defined on* $[0, 1]$ *which converges in measure to a real-valued function* f *defined on* $[0, 1]$*, then* $\{f_n\}_{n\in\mathbb{N}}$ *converges a.e. to* f.

Proof Observe that each f_n and f are measurable. Let

$$A = \{x \in (0, 1) \colon f \text{ is approximately continuous at } x\}.$$

From Theorem 3.13 it follows that $\lambda(A) = 1$. We shall show that $\lim_{n\to\infty} f_n(x) = f(x)$ for each $x \in A$. Suppose that this is not the case. Then there exist $x_0 \in A$, $\varepsilon > 0$ and an increasing sequence $\{n_m\}_{m\in\mathbb{N}}$ of positive integers such that $f_{n_m}(x_0) > f(x_0) + \varepsilon$ for each $m \in \mathbb{N}$ or $f_{n_m}(x_0) < f(x_0) - \varepsilon$ for each $m \in \mathbb{N}$. Consider the first case (in the second the reasoning is similar). There exist a measurable set $E \subset (0, 1)$ and a positive number δ such that $\lambda(E \cap (x_0, x_0 + \delta) > 0$ and $f(x) \in \left(f(x_0) - \frac{\varepsilon}{2}\right)$ for $x \in E \cap (x_0, x_0 + \delta)$. It is the consequence of the fact that f is approximately continuous at x_0. Then we have $f(x) < f(x_0) + \frac{\varepsilon}{2}$ and $f(x_0) + \varepsilon < f_{n_m}(x_0) \le f_{n_m}(x)$, so $f_{n_m}(x) - f(x) > \frac{\varepsilon}{2}$ for each $x \in E \cap (x_0, x_0 + \delta)$ and each $m \in \mathbb{N}$. It means that $\{f_n\}_{n\in\mathbb{N}}$ does not converge in measure to f—a contradiction. □

Remark 15.32 The case when f_n's are nonincreasing or the sequence $\{f_n\}_{n\in\mathbb{N}}$ consists of nondecreasing and nonincreasing functions can be treated similarly.

Remark 15.33 If we additionally assume that f is monotone, then $\lim_{n\to\infty} f_n(x) = f(x)$ at each point of continuity of f, so except on a denumerable set.

Consider now the space $\mathfrak{M}$ of equivalence classes of real-valued measurable functions defined on $[0, 1]$ with the distance function

$$d\left([f], [g]\right) = \int_0^1 \frac{|f(x) - g(x)|}{1 + |f(x) - g(x)|}\,\mathrm{d}x$$

(see Chap. 2 or Definition 0.8 and Remark 0.9).

Theorem 15.34 *If* $X \subset \mathfrak{M}$ *is open, convex and non-empty, then* $X = \mathfrak{M}$.

Proof We can assume that there exists $\varepsilon > 0$ such that $A = \{[g] \colon d([g], 0) < \varepsilon\} \subset X$. If $[f] \in \mathfrak{M}$ and $j \in \mathbb{N}$, then we put

$$f_{n,j}(x) = \begin{cases} n \cdot f(x) & \text{for } \frac{j-1}{n} \le x < \frac{j}{n} \\ 0 & \text{for remaining } x \end{cases}$$

for $j = 1, 2, \ldots, n$. Obviously $[f] = \sum_{j=1}^{n} \frac{1}{n}[f_{n_j,j}]$. Observe that $d\left([f_{n,j}], 0\right) \le \frac{1}{n}$ for $j = 1, 2, \ldots, n$ and $n \in \mathbb{N}$, so there exists $n_0 \in \mathbb{N}$ for which $f_{n_0,j} \in A$ for $j = 1, 2, \ldots, n$. Then $f \in A \subset X$, because X is convex and $X = \mathfrak{M}$. □

Remark 15.35 From this theorem it follows that there does not exist any norm describing the convergence in measure, because in the linear normed space there are open non-empty convex sets different from the whole space, for example $\{x : \| x \| < 1\}$. Recall that the "norm" in Chap. 2 (Theorem 2.9) does not fulfill the condition $\| af \| = |a| \cdot \| f \|$.

The convergence in category cannot be described with the use of any distance function (see Theorem 1.14). However, if $\mathfrak{B}$ is a space consisting of equivalence classes of real-valued functions defined on $[0, 1]$ having the Baire property (the functions f and g are equivalent, if $\{x \in [0, 1]\colon f(x) \neq g(x)\} \in \mathcal{K}$) equipped with the convergence in category and the topology in $\mathfrak{B}$ is defined below, then the situation is similar.

Definition 15.36 We say that a set $F \subset \mathfrak{B}$ is closed if the following condition is fulfilled; if a sequence $\{[f_n]\}_{n \in \mathbb{N}}$ of elements of F converges in category to $[f]$, then $[f] \in F$. The set $G \subset \mathfrak{B}$ is open if $\mathfrak{B} \setminus G$ is closed.

It is easy to verify that the family of open sets defined in this way is indeed a topology.

Remark 15.37 If we define the closure of the set $A \subset \mathfrak{B}$ in the following way:

$$\overline{A} = \bigcap \{F\colon F \text{ is closed and } F \supset A\},$$

Then we obtain the closure operation fulfilling all axioms of Kuratowski (see [Eng89]). It is clear that this closure is different from that in Theorem 2.14.

Theorem 15.38 *If $X \subset \mathfrak{B}$ is open, convex and nonempty, then $X = \mathfrak{B}$.*

Proof We can assume that $0 \in X$. If $[f] \in \mathfrak{B}$, then $f_{n,j}$ is defined exactly as in the proof of Theorem 15.34. We shall show that there exists $n_0 \in \mathbb{N}$ such that $[f_{n_0,j}] \in X$ for each $j = 1, 2, \ldots, n$. Suppose that there is not the case. hence for each $n \in \mathbb{N}$ there exists j_n, $1 \le j_n \le n$, for which $[f_{n,j_n}] \in \mathfrak{B} \setminus X$. Observe that the sequence $\left\{[f_{n,j_n}]\right\}_{n \in \mathbb{N}}$ converges in category to 0. Since $\mathfrak{B} \setminus X$ is closed, $0 \in \mathfrak{B} \setminus X$ – a contradiction. it follows that $[f] = \sum_{j=1}^{n} \frac{1}{n_0}[f_{n_0,j}]$ belongs to X and $X = \mathfrak{B}$. □

Remark 15.39 It can be proved that there is a nonzero continuous linear functional on a linear topological space if and only if this space properly contains a convex open non-empty set [KN63] Hence Theorems 15.34 and 15.40 follow immediately from Theorem 15.11 (and conversely)

Below we shall always assume that $\{a_n\}_{n \in \mathbb{N}}$ is a sequence of positive numbers such that $\sum_{n=1}^{\infty} a_n = +\infty$.

Definition 15.40 We shall say that $\sum_{k=1}^{\infty} a_{n_k}$ is a subseries of the series $\sum_{n=1}^{\infty} a_n = +\infty$ when $\{n_k\}_{k \in \mathbb{N}}$ is an increasing sequence of positive integers.

If $x \in (0, 1]$, then x can be represented uniquely by means of infinite dyadic expansion $x = \sum_{n=1}^{\infty} d_n(x) \cdot 2^{-n}$, where $d_n(x) = 0$ or 1 and $d_n(x) = 1$ for an infinite number of n's. If we put $S(x) = \sum_{n=1}^{\infty} d_n(x) \cdot a_n$ for $x \in (0, 1]$, then the set $\{S(x) \colon x \in (0, 1]\}$ can be treated as the set of all subseries of $\sum_{n=1}^{\infty} a_n$. Let

$$C\left(\sum_{n=1}^{\infty} a_n\right) = \{x \in (0, 1] \colon S(x) \text{ converges}\}$$

and

$$D\left(\sum_{n=1}^{\infty} a_n\right) = \{x \in (0, 1] \colon S(x) \text{ diverges}\}.$$

We shall present properties of $C\left(\sum_{n=1}^{\infty} a_n\right)$ and $D\left(\sum_{n=1}^{\infty} a_n\right)$ connected with measure and category. Much more information one can find in [Šal64, Šal68, Aue30].

Lemma 15.41 *$C\left(\sum_{n=1}^{\infty} a_n\right)$ is measurable and has the Baire property.*

Proof Put $S_n(x) = \sum_{k=1}^{n} d_k(x) \cdot a_k$ for $n \in \mathbb{N}$. Then

$$C\left(\sum_{n=1}^{\infty} a_n\right) = \bigcap_{p=1}^{\infty} \bigcup_{n=1}^{\infty} \bigcap_{k=n}^{\infty} \bigcap_{m=n}^{\infty} \left\{x \in (0, 1] \colon |S_k(x) - S_m(x)| < \frac{1}{p}\right\},$$

so $C\left(\sum_{n=1}^{\infty} a_n\right) \in \mathcal{L} \cap \mathcal{B}$, because all functions d_n (and also S_n) are measurable and have the Baire property. □

Lemma 15.42 *For each interval $(a, b) \subset (0, 1]$ and each positive dyadic rational h such that $(a + h, b + h) \subset (0, 1]$ we have*

$$\lambda\left((a, b) \cap C\left(\sum_{n=1}^{\infty} a_n\right)\right) = \lambda\left((a + h, b + h) \cap C\left(\sum_{n=1}^{\infty} a_n\right)\right).$$

Proof Let $N_n(x) = \{n \in \mathbb{N} \colon d_n(x) = 1\}$. If h is a positive dyadic rational, i.e. $h = \sum_{n=1}^{n_0} d_n(x) \cdot 2^{-n}$ for some $n_0 \in \mathbb{N}$, then $\operatorname{card}(N(x) \triangle N(x + h)) < \infty$. Hence $x \in (a, b) \cap C\left(\sum_{n=1}^{\infty} a_n\right)$ if and only if $x + h \in (a + h, b + h) \cap C\left(\sum_{n=1}^{\infty} a_n\right)$ and the thesis follows from the invariance of Lebesgue measure under translation. □

For the convenience we shall prove the following well known result.

Lemma 15.43 *If $E \subset (0, 1]$ is a measurable set such that $\lambda((a, b) \cap E) = \lambda((a + h, b + h) \cap E)$ for each $(a, b) \subset (0, 1]$ and each positive dyadic rational h such that $(a + h, b + h) \subset (0, 1]$, then $\lambda(E) = 1$ or $\lambda(E) = 1$.*

Proof Suppose that this is not the case, so $0 < \lambda(E) < 1$. then from the Lebesgue Density Theorem (see p. 42) it follows that there exist points $x_1, x_2 \in (0, 1)$ such that

x_1 is a density point of E and x_2 is a dispersion point of E. Suppose for the moment that $x_1 < x_2$ (if $x_2 < x_1$ the reasoning is the same). There exists $\delta > 0$ such that for each interval (a_1, b_1) such that $x_1 \in (a_1, b_1)$ and $b_1 - a_1 < \delta$ we have $\frac{\lambda((a_1,b_1)\cap E)}{b_1-a_1} > 0.9$ and for each interval (a_2, b_2) such that $x_2 \in (a_2, b_2)$ and $b_2 - a_2 < \delta$ we have $\frac{\lambda((a_2,b_2)\cap E)}{b_2-a_2} < 0.1$. It is not difficult to observe that (a_1, b_1) and (a_2, b_2) can be found such that $(b_1 - a_1) = (b_2 - a_2) < \delta$ and $a_2 - a_1$ is a positive dyadic rational. This contradicts the equality in Lemma 15.42. □

Theorem 15.44 *([Šal68])* $\lambda\left(C\left(\sum_{n=1}^{\infty} a_n\right)\right) = 0$ *and* $\lambda\left(D\left(\sum_{n=1}^{\infty} a_n\right)\right) = 1$.

Proof From Lemmas 15.42 and 15.43 it follows that either $\lambda\left(C\left(\sum_{n=1}^{\infty} a_n\right)\right) = 0$ or $\lambda\left(C\left(\sum_{n=1}^{\infty} a_n\right)\right) = 1$. Suppose that $\lambda\left(C\left(\sum_{n=1}^{\infty} a_n\right)\right) = 1$. Then there exists $x_0 \in C\left(\sum_{n=1}^{\infty} a_n\right)$ such that $1 - x_0 \in C\left(\sum_{n=1}^{\infty} a_n\right)$. Observe that x_0 has infinitely many 0s and infinitely many 1s in its dyadic expansion and similarly behaves $1 - x_0$. We have $\sum_{n=1}^{\infty} d_n(x_0)a_n < \infty$ and $\sum_{n=1}^{\infty} d_n(1 - x_0)a_n < \infty$, so $\sum_{n=1}^{\infty} a_n = \sum_{n=1}^{\infty} d_n(x_0)a_n + \sum_{n=1}^{\infty} d_n(1 - x_0)a_n < \infty$, which contradicts our assumption. Hence $\lambda\left(C\left(\sum_{n=1}^{\infty} a_n\right)\right) = 0$ and $\lambda\left(D\left(\sum_{n=1}^{\infty} a_n\right)\right) = 1 - \lambda\left(C\left(\sum_{n=1}^{\infty} a_n\right)\right) = 1$. □

The above theorem has a satisfactory counterpart for category:

Theorem 15.45 *([Šal68])* $C\left(\sum_{n=1}^{\infty} a_n\right) \in \mathcal{K}$ *and* $D\left(\sum_{n=1}^{\infty} a_n\right)$ *is residual in* $(0, 1]$.

Proof Consider the space $X = (0, 1] \setminus \left\{\frac{k}{2^n} : n \in \mathbb{N} \text{ and } 1 \leq k \leq 2^n\right\}$ equipped with the euclidean metric. Put $S_n(x) = \sum_{k=1}^{n} d_k(x) \cdot a_k$ for $n \in \mathbb{N}$, as in Lemma 15.41. Each S_n is continuous on X, so the set $E(p) = \bigcap_{n=1}^{\infty} \{x \in X : S_n(x) \leq p\}$ is closed in X for each $p \in \mathbb{N}$. We shall show that each $E(p)$ is nowhere dense in X. Let $(a, b) \subset (0, 1]$ be an arbitrary interval. It suffices to prove the existence of $x_0 \in \big((a, b) \cap X\big) \setminus E(p)$. It follows immediately from Theorem 15.44 since $\lambda\Big(D\left(\sum_{n=1}^{\infty} a_n\right) \cap (a, b)\Big) = b - a$. Then $C\left(\sum_{n=1}^{\infty} a_n\right) = \bigcup_{p=1}^{\infty} E(p)$ is of the first category in X and thus also of the first category in $(0, 1]$. The second part of thesis is obvious. □

References

[AU23] P.S. Alexandroff, P.S. Urysohn, Une condition nécessaire et suffisante pour qu'une classe (L) soit une classe (D). C. R. Acad. Sci. **177**, 1274–1276 (1923)

[Aue30] H. Auerbach, Uber die Vorzeichenverteilung in unendlichen Reichen. Studia Math. **2**, 228–230 (1930)

[BK29] S. Banach, K. Kuratowski, Sur une généralisation du problème de la mesure. Fund. Math. **13**, 127–131 (1929)

[Bari61] N.K. Bari, *Trigonometriceskije riady* (Moscow, 1961) (in Russian)

[Bar80] R.G. Bartle, An extension of Egorov's theorem. Am. Math. Mon. **87**(8), 628–633 (1980)

[Bee81] G. Beer, A natural topology for upper semicontinuous function and a Baire category dual for convergence in measure. Pac. J. Math. **96**(2), 251–263 (1981)

[Bee82] G. Beer, Upper semicontinuous functions and the stone approximation theorem. J. Approx. Theory **34**, 1–11 (1982)

[BKL86] R.G. Bilyeu, R.R. Kallman, P.W. Lewis, Rearrangements and category. Pac. J. Math. **121**(1), 41–46 (1986)

[Blu35] H. Blumberg, The measurable boundaries of an arbitrary function. Acta Math. **65**, 263–282 (1935)

[BD78] D. Borwein, S.Z. Ditor, Translates of sequences in sets of positive measure. Can. Math. Bull. **21**(4), 497–498 (1978)

[BFT78] J. Bourgain, D.H. Fremlin, M. Talagrand, Pointwise compact sets of Baire-measurable functions. Am. J. Math. **100**(4), 845–886 (1978)

[BG78] D. Bucchioni, A. Goldman, Sur la convergence presque partout des suites de fonctions mesurables. Can. J. Math. (3), 638–649 (1978)

[Bru73] A. Bruckner, The differentiability properties of typical functions in $C[a, b]$. Am. Math. Monthly **80**, 679–683 (1973)

[BBT97] A.M. Bruckner, J.B. Bruckner, B.S. Thomson, *Real Analysis* (Prentice Hall, 1997)

[BG77] A. Bruckner, K. Garg, The level structure of a residual set of continuous functions. Trans. Am. Math. Soc. **232**, 307–321 (1977)

[BP43] R.C. Buck, H. Pollard, Convergence and summability properties of subsequences. Bull. Am. Math. Soc. **49**, 924–931 (1943)

[Car66] L. Carleson, On convergence and growth of partial sums of Fourier series. Acta Math. **116**, 135–157 (1966)

[CLO91] K. Ciesielski, L. Larson, K. Ostaszewski, $\mathcal{I}$-density continuous functions. Mem. Am. Math. Soc. (1991)

W. Wilczyński, *Convergence in Measure and in Category*, Lecture Notes in Mathematics 2394, https://doi.org/10.1007/978-3-032-23143-7

[Den15] A. Denjoy, Mémoire sur les dérivés des fonctions continues. J. Mat. Pures et Appl. **1**, 105–240 (1915)
[Ego11] D.T. Egorov, Sur les séries des fonctions mesurables. C. R. Paris **152**, 226–253 (1911)
[Eng89] R. Engelking, *General Topology* (Heldermann Verlag, 1989)
[Erd43] P. Erdös, Some remarks on the set theory. Ann. Math. **44**(2), 643–646 (1943)
[Frt27] M. Frechet, Sur les ensembles compacts de fonctions mesurable. Fund. Math. **9**, 25–32 (1927)
[Fre75] D.H. Fremlin, Pointwise compact sets of measurable functions. Manuscripta Math. **15**, 219–242 (1975)
[Gar64] A.M. Garsia, Existence of almost everywhere convergent rearragements for Fourier series of L_2 functions. Ann. Math. **79**(3), 623–629 (1964)
[GNN61] C. Goffman, C.J. Neugebauer, T. Nishiura, The density topology and approximate continuity. Duke Math. J. **28**, 497–505 (1961)
[GW60] C. Goffman, D. Waterman, Basic sequences in the space of measurable functions. Proc. Am. Math. Soc. **11**, 211–213 (1960)
[GW60a] C. Goffman, D. Waterman, On upper and lower limits in measure. Fund. Math. XLVIII. **2**, 128–133 (1960)
[GW61] C. Goffman, D. Waterman, Approximately continuous transformations. Proc. Am. Math. Soc. **12**, 116–121 (1961)
[GZ60] C. Goffman, R.E. Zink, Concerning the measurable boundaries of a real function. Fund. Math. XLVIII. **2**, 105–111 (1960)
[GH55] W.H. Gottschalk, G.A. Hedlund, *Topological Dynamics*, vol. XXXVI (AMS Colloquium Publications, Providence, R. I., 1955)
[Gri66] J.M. Gribanov, Remark on convergence a.e. and convergence in measure. Comm. Math. Univ. Carolinae **7**(3), 297–300 (1966) (in Russian)
[G-E99] K.-G. Grosse-Erdman, Universal families and hypercyclic operators. Bull. Am. Math. Soc. **36**(3), 345–381 (1999)
[Hal65] P.R. Halmos, *Measure Theory* (Van Nostrand, Tenth printing, 1965)
[HL14] G.H. Hardy, J.E. Littlewood, Some problems of Diophantine Approximations. Acta Math. **37**, 155–239 (1914)
[HWr62] G.H. Hardy, E.M. Wright, *An Introduction to the Theory of Numbers* (Oxford at the Clarendon Press, London, 1962)
[HP52] O. Haupt, Ch. Pauc, La topologie de Denjoy envisagée vraie topologie. C. R. Acad. Sci. Paris **234**, 390–392 (1952)
[Hau62] F. Hausdorff, *Set Theory* (Chelsea Publishing Company, New York, 1962)
[HWa91] J. Hejduk, E. Wajch, A characterization of compactness in the sense of convergence with respect to a σ-ideal. Radovi Matematicky **7**, 5–9 (1991)
[HS69] E. Hewitt, K. Stromberg, *Real and Abstract Analysis* (Springer Verlag, 1969)
[Jec78] T. Jech, *Set Theory* (Academic, 1978)
[Kah00] J.P. Kahane, Baire's category theorem and trigonometric series. Journal d'Analyse Mathematique **80**, 143–188 (2000)
[Kan30] L. Kantorovich, Sur les suites des fonctions presque partout continues. Fund. Math. **17**, 25–28 (1930)
[KS84] B.S. Kashin, A.A. Saakjan, *Orthogonal Series (Russian)* (Science, Moscow, 1984)
[KN63] J.L. Kelley, I. Namioka, *Linear Topological Spaces* (Van Nostrand, 1963)
[Kno51] K. Knopp, *Theory and Application of Infinite Series* (Blackie and Son, London, 1951)
[Kol24] A.N. Kolmogorov, Une contribution à l'étude de la convergence des series de Fourier. Fund. Math. **5**, 96–97 (1924)
[Kur24] K. Kuratowski, Sur les fonctions représentable analitiquement et les ensembles da premiere categorie. Fund. Math. **5**, 75–86 (1924)
[Kur66] K. Kuratowski, *Topology*, vol. 1 (Academic, 1966)
[Leb05] H. Lebesgue, Sur les fonctions représentable analytiquement. J. Math (6), 1 (1905)
[Lus12] N.N. Lusin, Sur les propriétés des fonctions mesurables. C. R. **154**, 1688–1690 (1912)

[Lus34] N.N. Lusin, On sequences of measurable function, in *Integral and Trigonometriceskij Riad.* ed. by N.K. Bari, D.E. Menchov (Moskwa-Leningrad, 1951), pp. 327–337 (in Russian)

[LS18] N.N. Lusin, W. Sierpiński, Sur quelques propriétés des ensembles (A). Bull. Int. Acad. Sci. Cracovie A (1918), 35–48 (OC, Vol. 2, 192–204)

[Marc33] J. Marcinkiewicz, A new proof of the theorem on Fourier series. J. Lond. Math. Soc. **8**, 179 (1933)

[Mar55] E. Marczewski, Remarks on the convergence of measurable sets and measurable functions. Coll. Math. **3**, 118–124 (1955)

[Mau73a] R.D. Mauldin, On Borel measures and Baire class 3. Proc. Am. Math. Soc. **39**, 308–312 (1973)

[Mau73b] R.D. Mauldin, The Baire order of the functions continuous almost everywhere. Proc. Am. Math. Soc. **41**(2), 535–540 (1973)

[MO40] S. Mazur, W. Orlicz, Sur quelques propriétés de fonctions périodiques et presque-periodiques. Studia Math. **9**, 1–16 (1940)

[Mei73] A. Meir, Solution of problem 5836. Am. Math. Monthly **80**(IV.4), 444–445 (1973)

[Mil96] A.W. Miller, Soulsin's hypothesis and convergence in category. Proc. Am. Math. Soc. **124**(5), 1529–1532 (1996)

[Mi85] H.J. Miller, On certain transformations of sets. Akad. Nauka Umjet. Bosne Hercegov. Rad. Odjelj. Prirod. Mat. Nauka **24**, 5–9 (1985)

[Mi89] H.J. Miller, Generalization of a result of Borwein and Ditor. Proc. Am. Math. Soc. **105**(4), 889–893 (1989)

[MS96] H.J. Miller, F.J. Schnitzer, Measure and category—some non-analogues. Math. Pannonica **7**(1), 69–79 (1996)

[Nik25] O. Nikodym, Sur une propriété de l'operation (A). Fund. Math. **7**, 149–154 (1925)

[Orl48] W. Orlicz, Sur quelques propriétes des fonctions de Baire périodiques. Studia Math. **10**, 148–158 (1948)

[Orl51] W. Orlicz, On a class of asymptotically divergent sequences of functions, Studia Math. 12, *286–307; Collected papers, part I, 630–651* (Polish Scientific Publishers, Warszawa, 1951), p. 1988

[Oxt80] J.C. Oxtoby, *Measure and Category*, 2nd edn. (Springer, 1980)

[Pa14] J. Pál, Über eine Anwendung des Weierstrass-schen Satzes von der Annäherung stetiger Funktionen durch Polynome. Tohoku Math J. **5**, 8–9 (1914)

[PWW85] W. Poreda, E. Wagner-Bojakowska, W. Wilczyński, A category analogue of the density topology. Fundam. Math. **125**, 167–173 (1985)

[Ra22] H. Rademacher, Einige Sätze über reichen von allemeigne Orthogonalfunktionen. Math. Ann. **87**, 112 (1922)

[RW94] J. Rzepecka, W. Wilczyński, On the transformation of measurable sets and sets with the Baire property. RAE **20**(1), 178–182 (1994/1995)

[Ros77] H.P. Rosenthal, Point-wise compact subsets of the first Baire class. Am. J. Math. **99**(2), 362–378 (1977)

[Sak37] S. Saks, *Theory of the Integral* (Monografje Matematyczne, Tom VII, Warszawa-Lw w, 1937)

[Saz65] V. V. Sazonov, On perfect measures. Am. Math. Soc. Translations **48**(2), 229–254 (1965)

[SS28] S. Saks, W. Sierpiński, Sur une propriété générale de fonctions. Fund. Math. **11**, 105–112 (1928) (OC, Vol. 2, 678–684)

[Sik59] R. Sikorski, *Funkcje Rzeczywiste* (Tom II, PWN Warszawa, 1959). (in Polish)

[Sie17] W. Sierpiński, Sur un ensemble non mesurable. Tokohu Math. J. **12**, 205–208 (1917) (OC, Vol. 2, 183–186)

[Sie27] W. Sierpiński, Remarque sur la convergence en mesure. Fund. Math. **9**, 33–36 (1927) (OC, Vol. 2, 621–623)

[Sie28] W. Sierpiński, Remarque sur le thèoreme de M. Egoroff. C. R. Soc. Sci. et Lettr. Varsovie, Cl. III **21**, 84–87 (1928) (O. C. Vol. II, 667–670)

[Sie30] W. Sierpiński, Sur l'extension des fonctions de Baire définies sur les ensembles quelconques. Fund. Math. **16**, 81–89 (1930) (OC, Vol. 3, 36–43)

[Sie34] W. Sierpiński, Sur la dualité entre la premiére catégorie et la messure nulle. Fund. Math. **22**, 276–280 (1934)

[Sie45] W. Sierpiński, Sur une suite infinie de fonctions de classe 1 dont toute fonction d'accumulation est non mesurable. Fund. Math. **33**, 104–105 (1945) (OC, Vol. 3, 425–426)

[Sie47] W. Sierpiński, Sur une suite infinie de fonctions continues dont toute fonction d'accumulation est non mesurable. Publ Inst. Math. Beograd **1**, 5–10 (1947) (OC, Vol. 3, 553–557)

[Sie50] W. Sierpiński, Sur les suites doubles de fonctions. Fund. Math. **37**, 55–62 (1950)

[Sth20] H. Steinhaus, Uogólnienie pewnego twierdzenia G. Cantora. Wiadomości Matematyczne **24**, 197–201 (1920) (in Polish) (A generalization of Cantor's theorem on trigonometric series, Selected Papers 289–292)

[Sth28] H. Steinhaus, Sur une question concernant la convergence de séries des fonctions. Fund. Math. **11**, 186–192. Selected papers, Polish Scientific Publishers 1985, 373–379 (1928)

[S-N64] B. Szökefalvi-Nagy, *Introduction to Real Functions and Orthogonal Expansions* (Akadémia Kiadó, Budapest, 1964)

[Szp33] E. Szpilrajn (Marczewski), Sur certains invariants de l'operation (A). Fund. Math. **21**, 229–235 (1933) (Collected Papers 50–56)

[Szp34] E. Szpilrajn, Remarques sur les fonctions comlètement additives d'essemble et sur les ensembles joissant de la propriété de Baire. Fund. Math. **22**, 303–311 (1934)

[Šal64] T. Šalát, On subseries. Math. Zeitschr. **85**, 209–225 (1964)

[Šal68] T. Šalát, On subseries of divergent series. Matematický časopis **18**(4), 312–338 (1968)

[Ta-d80] M. Talagrand, Compact de fonctions mesurables et filtres non measurables. Stud. Math. **47**, 13–43 (1980)

[Ta-d84] M. Talagrand, Pettis integral and measure theory. Mem. Am. Math. Soc. **307** (1984)

[Ta-n57] A.A. Talalyan, On the convergence almost everywhere of subsequences of partial sums of general orthogonal series. Akad. Nauk Armyan SSR. Izv. Fiz.-Mat. Estest. Tehn. Nauki **10**, 17–34 (Russian, Armenian summary, 1957)

[Tay60] S.J. Taylor, An alternative form of Egoroff's theorem. Fund. Math. **48**(2), 169–174 (1959)

[Tod84] S. Todorčevič, *Trees and Linearly Ordered Sets in Handbook of Set Theoretic Topology* (North Holland, 1984), pp. 235–293

[Tre60] N.I. Trebukova, Metric convergence and metric isomorphism. Uspekhi Mat. Nauk **15**(2(92)), 195–199 (1960) (in Russian)

[Tyc35] A. Tychonoff, Über einen Funktionchraum. Math. Annalen **111**, 762 (1935)

[Vai64] R. Vaidyanathaswamy, *Set Topology* (Chelsea Publishing Company, New York, 1964)

[Vul67] B.Z. Vulikh, *Introduction to the Theory of Partially Ordered Sets* (Wolters-Noordhoff Scientific Publications, 1967)

[Vul76] B.Z. Vulikh, *A Brief Course in the Theory of Functions of a Real Variable* (Mir Publishers, Moscow, 1976)

[Wag81] E. Wagner, Sequences of measurable functions. Fund. Math. CXI **I**, 89–102 (1981)

[W-BW80] E. Wagner-Bojakowska, W. Wilczyński, Convergence of sequences of measurable functions. Acta Math. Acad. Sci. Hungaricae **36**(1–2), 125–128 (1980)

[W-BW81] E. Wagner-Bojakowska, W. Wilczyński, Convergence almost everywhere of sequences of measurable functions. Coll. Math. **XLV** (1981), Fasc. 1, 119–124

[W-BW99] E. Wagner-Bojakowska, W. Wilczyński, Cauchy condition for the convergence in category. Proc. Am. Math. Soc. **128**(2), 413–418 (1999)

[Wal77] W. Walter, A counterexample in connection with Egorov's theorem. Am. Math. Monthly **84**(2), 118–119 (1977)

[Wil02] W. Wilczyński, Density topologies, in *Chapter 15 in Handbook of Measure Theory*. ed. by E. Pap (Elsevier, 2002), pp.675–702

[Wil16] W. Wilczyński, Convergence in measure and in category, similarities and differences. Aequationes Math. **90**, 99–105 (2016)

[Wil19] W. Wilczyński, On series of functions with the Baire property. Eur. J. Math. **5**(1), 168–173 (2019)

[Wil20] W. Wilczyński, Convergence in measure and in category. Ann. Math. Sil. **34**(1), 164–168 (2020)

[WK92] W. Wilczyński, A. Kharazishvili, On the translations of measurable sets and sets with the Baire property. Bull. of the Acad. Sci. Georgia **145**(1), 43–46 (1992) (in Russian; English and Georgian Summary)

[WW05] W. Wilczyński, W. Wojdowski, Upper and lowr limits of sequences of measurable sets and of sets with the Baire Property. Acta. Univ. Lodziensis, Folia Mathematica **12**(1), 73–80 (2005)

[Wla60] D.A. Wladimirov, *On the completeness of a partially ordered space*. Uspekhi Mat. Nauk **15**(2(92)), 165–172 (1960) (in Russian)

[Yon75] K. Yoneda, On control function of a.e. convergence. Math. Jpn. **20**, 101–105 (1975)

[You16] W.H. Young, Note on functions of upper and lower type. Proc. Lond. Math. Soc. **15**, 354–359 (1916)

[Zak64] E. Zakon, A remark on the theorems of Lusin and Egoroff. Can. Math. Bull. **7**(2), 291–295 (1964)

[Zal23] Z. Zalcwasser, in Problèmes. Fund. Math. **4**, 368–370 (1923)

LECTURE NOTES IN MATHEMATICS

Springer

Editors in Chief: J.-M. Morel, B. Teissier

Editorial Policy

1. Lecture Notes aim to report new developments in all areas of mathematics and their applications – quickly, informally and at a high level. Mathematical texts analysing new developments in modelling and numerical simulation are welcome. See Lecture Notes in Mathematics for the existing online volumes of LNM.

   Manuscripts should be reasonably self-contained and rounded off. Thus, they may, and often will, present not only results of the author but also related work by other people. They may be based on specialised lecture courses. Furthermore, the manuscripts should provide sufficient motivation, examples and applications. This clearly distinguishes Lecture Notes from journal articles or technical reports which normally are very concise. Articles intended for a journal but too long to be accepted by most journals, usually do not have this "lecture notes" character. For similar reasons it is unusual for doctoral theses to be accepted for the Lecture Notes series, though habilitation theses may be appropriate.

2. Besides monographs, multi-author manuscripts resulting from SUMMER SCHOOLS or similar INTENSIVE COURSES are welcome, provided their objective was held to present an active mathematical topic to an audience at the beginning or intermediate graduate level (a list of participants should be provided).

   The resulting manuscript should not be just a collection of course notes but should require advance planning and coordination among the main lecturers. The subject matter should dictate the structure of the book. This structure should be motivated and explained in a scientific introduction, and the notation, references, index and formulation of results should be, if possible, unified by the editors. Each contribution should have an abstract and an introduction referring to the other contributions. In other words, more preparatory work must go into a multi-authored volume than simply assembling a disparate collection of papers, communicated at the event.

3. Manuscripts should be submitted online either at www.editorialmanager.com/lnm to Springer's mathematics editorial in Heidelberg, or electronically to one of the series editors. Authors should be aware that incomplete or insufficiently close-to-final manuscripts almost always result in longer refereeing times and nevertheless unclear referees' recommendations, making further refereeing of a final draft necessary. The strict minimum amount of material that will be considered should include a detailed outline describing the planned contents of each chapter, a bibliography and several sample chapters. Parallel submission of a manuscript to another publisher while under consideration for LNM is not acceptable and can lead to rejection.

4. The type of material considered for publication includes:

   - Research monographs
   - Lectures on a new field or presentations of a new angle in a classical field
   - Summer schools and intensive courses on topics of current research

In general, **monographs** will be sent out to at least 2 external referees for evaluation.

A final decision to publish can be made only on the basis of the complete manuscript, however a refereeing process leading to a preliminary decision can be based on a pre-final or incomplete manuscript.

Volume Editors of **multi-author works** are expected to arrange for the refereeing, to the usual scientific standards, of the individual contributions. If the resulting reports can be forwarded to the LNM Editorial Board, this is very helpful. If no reports are forwarded or if other questions remain unclear in respect of homogeneity etc., the series editors may wish to consult external referees for an overall evaluation of the volume.

5. Manuscripts should in general be submitted in English. Final manuscripts should contain at least 100 pages of mathematical text and should always include:
   - A table of contents
   - An informative introduction, with adequate motivation and perhaps some historical remarks: it should be accessible to a reader not intimately familiar with the topic treated
   - A subject index: as a rule, this is genuinely helpful for the reader
   - For evaluation purposes, manuscripts should be submitted as pdf files.

6. Careful preparation of the manuscripts will help keep production time short besides ensuring satisfactory appearance of the finished book in print and online. After acceptance of the manuscript authors will be asked to prepare the final LaTeX source files plus the corresponding pdf- or zipped ps-file (Preparing your Manuscript). You can also download our LaTeX package for monographs. The LaTeX source files are essential for producing the full-text online version of the book. For more detailed information, please visit the Book Manuscript Guidelines and Key Style Points. The technical production of a Lecture Notes volume takes approximately 12 weeks. Additional instructions, if necessary, are available on request from lnm@springer.com.

7. Authors receive a total of 30 free copies of their volume and free access to their book on SpringerLink, but no royalties. They are entitled to a discount of 40 % on the price of Springer books purchased for their personal use, if ordering directly from Springer.

8. Commitment to publish is made by a *Publishing Agreement*; contributing authors of multi-author books are requested to sign a *Consent to Publish form*. Springer-Verlag registers the copyright for each volume. Authors are free to reuse material contained in their LNM volumes in later publications: a brief written (or e-mail) request for formal permission is sufficient.

Addresses:

Professor Jean-Michel Morel, City University Hongkong, Kowloon Tong, Hong Kong
E-mail: moreljlmichel@gmail.com

Professor Bernard Teissier, Equipe Géométrie et Dynamique,
Institut de Mathématiques de Jussieu – Paris Rive Gauche, Paris, France
E-mail: bernard.teissier@imj-prg.fr

Springer: Ute McCrory, Editor, Mathematics, Heidelberg, Germany.
E-mail: lnm@springer.com

Index

W. Wilczyński, *Convergence in Measure and in Category*, Lecture Notes in Mathematics 2394, https://doi.org/10.1007/978-3-032-23143-7